52 Wave propagation in viscoelastic media
F Mainardi

53 Nonlinear partial differential equations and their applications: Collège de France Seminar. Volume I
H Brezis and J L Lions

54 Geometry of Coxeter groups
H Hiller

55 Cusps of Gauss mappings
T Banchoff, T Gaffney and C McCrory

56 An approach to algebraic K-theory
A J Berrick

57 Convex analysis and optimization
J-P Aubin and R B Vintner

58 Convex analysis with applications in the differentiation of convex functions
J R Giles

59 Weak and variational methods for moving boundary problems
C M Elliott and J R Ockendon

60 Nonlinear partial differential equations and their applications: Collège de France Seminar. Volume II
H Brezis and J L Lions

61 Singular systems of differential equations II
S L Campbell

62 Rates of convergence in the central limit theorem
Peter Hall

63 Solution of differential equations by means of one-parameter groups
J M Hill

64 Hankel operators on Hilbert space
S C Power

65 Schrödinger-type operators with continuous spectra
M S P Eastham and H Kalf

66 Recent applications of generalized inverses
S L Campbell

67 Riesz and Fredholm theory in Banach algebra
B A Barnes, G J Murphy, M R F Smyth and T T West

68 Evolution equations and their applications
F Kappel and W Schappacher

69 Generalized solutions of Hamilton-Jacobi equations
P L Lions

70 Nonlinear partial differential equations and their applications: Collège de France Seminar. Volume III
H Brezis and J L Lions

71 Spectral theory and wave operators for the Schrödinger equation
A M Berthier

72 Approximation of Hilbert space operators I
D A Herrero

73 Vector valued Nevanlinna Theory
H J W Ziegler

74 Instability, nonexistence and weighted energy methods in fluid dynamics and related theories
B Straughan

75 Local bifurcation and symmetry
A Vanderbauwhede

76 Clifford analysis
F Brackx, R Delanghe and F Sommen

77 Nonlinear equivalence, reduction of PDEs to ODEs and fast convergent numerical methods
E E Rosinger

78 Free boundary problems, theory and applications. Volume I
A Fasano and M Primicerio

79 Free boundary problems, theory and applications. Volume II
A Fasano and M Primicerio

80 Symplectic geometry
A Crumeyrolle and J Grifone

81 An algorithmic analysis of a communication model with retransmission of flawed messages
D M Lucantoni

82 Geometric games and their applications
W H Ruckle

83 Additive groups of rings
S Feigelstock

84 Nonlinear partial differential equations and their applications: Collège de France Seminar. Volume IV
H Brezis and J L Lions

85 Multiplicative functionals on topological algebras
T Husain

86 Hamilton-Jacobi equations in Hilbert spaces
V Barbu and G Da Prato

87 Harmonic maps with symmetry, harmonic morphisms and deformations of metrics
P Baird

88 Similarity solutions of nonlinear partial differential equations
L Dresner

89 Contributions to nonlinear partial differential equations
C Bardos, A Damlamian, J I Díaz and J Hernández

90 Banach and Hilbert spaces of vector-valued functions
J Burbea and P Masani

91 Control and observation of neutral systems
D Salamon

92 Banach bundles, Banach modules and automorphisms of C*-algebras
M J Dupré and R M Gillette

93 Nonlinear partial differential equations and their applications: Collège de France Seminar. Volume V
H Brezis and J L Lions

94 Computer algebra in applied mathematics: an introduction to MACSYMA
R H Rand

95 Advances in nonlinear waves. Volume I
L Debnath

96 FC-groups
M J Tomkinson

97 Topics in relaxation and ellipsoidal methods
M Akgül

98 Analogue of the group algebra for topological semigroups
H Dzinotyiweyi

99 Stochastic functional differential equations
S E A Mohammed

100 Optimal control of variational inequalities
V Barbu
101 Partial differential equations and dynamical systems
W E Fitzgibbon III
102 Approximation of Hilbert space operators. Volume II
C Apostol, L A Fialkow, D A Herrero and D Voiculescu
103 Nondiscrete induction and iterative processes
V Ptak and F-A Potra
104 Analytic functions – growth aspects
O P Juneja and G P Kapoor
105 Theory of Tikhonov regularization for Fredholm equations of the first kind
C W Groetsch
106 Nonlinear partial differential equations and free boundaries. Volume I
J I Díaz
107 Tight and taut immersions of manifolds
T E Cecil and P J Ryan
108 A layering method for viscous, incompressible L_p flows occupying R^n
A Douglis and E B Fabes
109 Nonlinear partial differential equations and their applications: Collège de France Seminar. Volume VI
H Brezis and J L Lions
110 Finite generalized quadrangles
S E Payne and J A Thas
111 Advances in nonlinear waves. Volume II
L Debnath
112 Topics in several complex variables
E Ramírez de Arellano and D Sundararaman
113 Differential equations, flow invariance and applications
N H Pavel
114 Geometrical combinatorics
F C Holroyd and R J Wilson
115 Generators of strongly continuous semigroups
J A van Casteren
116 Growth of algebras and Gelfand–Kirillov dimension
G R Krause and T H Lenagan
117 Theory of bases and cones
P K Kamthan and M Gupta
118 Linear groups and permutations
A R Camina and E A Whelan
119 General Wiener–Hopf factorization methods
F-O Speck
120 Free boundary problems: applications and theory, Volume III
A Bossavit, A Damlamian and M Fremond
121 Free boundary problems: applications and theory, Volume IV
A Bossavit, A Damlamian and M Fremond
122 Nonlinear partial differential equations and their applications: Collège de France Seminar. Volume VII
H Brezis and J L Lions
123 Geometric methods in operator algebras
H Araki and E G Effros
124 Infinite dimensional analysis–stochastic processes
S Albeverio
125 Ennio de Giorgi Colloquium
P Krée
126 Almost-periodic functions in abstract spaces
S Zaidman
127 Nonlinear variational problems
A Marino, L Modica, S Spagnolo and M Degiovanni
128 Second-order systems of partial differential equations in the plane
L K Hua, W Lin and C-Q Wu
129 Asymptotics of high-order ordinary differential equations
R B Paris and A D Wood
130 Stochastic differential equations
R Wu
131 Differential geometry
L A Cordero
132 Nonlinear differential equations
J K Hale and P Martinez-Amores
133 Approximation theory and applications
S P Singh
134 Near-rings and their links with groups
J D P Meldrum
135 Estimating eigenvalues with *a posteriori/a priori* inequalities
J R Kuttler and V G Sigillito
136 Regular semigroups as extensions
F J Pastijn and M Petrich
137 Representations of rank one Lie groups
D H Collingwood
138 Fractional calculus
G F Roach and A C McBride
139 Hamilton's principle in continuum mechanics
A Bedford
140 Numerical analysis
D F Griffiths and G A Watson
141 Semigroups, theory and applications. Volume I
H Brezis, M G Crandall and F Kappel
142 Distribution theorems of L-functions
D Joyner
143 Recent developments in structured continua
D De Kee and P Kaloni
144 Functional analysis and two-point differential operators
J Locker
145 Numerical methods for partial differential equations
S I Hariharan and T H Moulden
146 Completely bounded maps and dilations
V I Paulsen
147 Harmonic analysis on the Heisenberg nilpotent Lie group
W Schempp
148 Contributions to modern calculus of variations
L Cesari
149 Nonlinear parabolic equations: qualitative properties of solutions
L Boccardo and A Tesei
150 From local times to global geometry, control and physics
K D Elworthy

Pitman Research Notes in Mathematics Series

Main Editors
H. Brezis, Université de Paris
R. G. Douglas, State University of New York at Stony Brook
A. Jeffrey, University of Newcastle-upon-Tyne *(Founding Editor)*

Editorial Board
R. Aris, University of Minnesota
A. Bensoussan, INRIA, France
S. Bloch, University of Chicago
B. Bollobás, University of Cambridge
W. Bürger, Universität Karlsruhe
S. Donaldson, University of Oxford
J. Douglas Jr, Purdue University
R. J. Elliott, University of Alberta
G. Fichera, Università di Roma
R. P. Gilbert, University of Delaware
R. Glowinski, Université de Paris
K. P. Hadeler, Universität Tübingen
K. Kirchgässner, Universität Stuttgart
B. Lawson, State University of New York at Stony Brook
W. F. Lucas, Claremont Graduate School
R. E. Meyer, University of Wisconsin-Madison
S. Mori, Nagoya University
L. E. Payne, Cornell University
G. F. Roach, University of Strathclyde
J. H. Seinfeld, California Institute of Technology
B. Simon, California Institute of Technology
I. N. Stewart, University of Warwick
S. J. Taylor, University of Virginia

Submission of proposals for consideration

Suggestions for publication, in the form of outlines and representative samples, are invited by the Editorial Board for assessment. Intending authors should approach one of the main editors or another member of the Editorial Board, citing the relevant AMS subject classifications. Alternatively, outlines may be sent directly to the publisher's offices. Refereeing is by members of the board and other mathematical authorities in the topic concerned, throughout the world.

Preparation of accepted manuscripts

On acceptance of a proposal, the publisher will supply full instructions for the preparation of manuscripts in a form suitable for direct photo-lithographic reproduction. Specially printed grid sheets are provided and a contribution is offered by the publisher towards the cost of typing. Word processor output, subject to the publisher's approval, is also acceptable.

Illustrations should be prepared by the authors, ready for direct reproduction without further improvement. The use of hand-drawn symbols should be avoided wherever possible, in order to maintain maximum clarity of the text.

The publisher will be pleased to give any guidance necessary during the preparation of a typescript, and will be happy to answer any queries.

Important note

In order to avoid later retyping, intending authors are strongly urged not to begin final preparation of a typescript before receiving the publisher's guidelines and special paper. In this way it is hoped to preserve the uniform appearance of the series.

Longman Scientific & Technical
Longman House
Burnt Mill
Harlow, Essex, UK
(tel (0279) 426721)

Titles in this series

151 A stochastic maximum principle for optimal control of diffusions
U G Haussmann
152 Semigroups, theory and applications. Volume II
H Brezis, M G Crandall and F Kappel
153 A general theory of integration in function spaces
P Muldowney
154 Oakland Conference on partial differential equations and applied mathematics
L R Bragg and J W Dettman
155 Contributions to nonlinear partial differential equations. Volume II
J I Díaz and P L Lions
156 Semigroups of linear operators: an introduction
A C McBride
157 Ordinary and partial differential equations
B D Sleeman and R J Jarvis
158 Hyperbolic equations
F Colombini and M K V Murthy
159 Linear topologies on a ring: an overview
J S Golan
160 Dynamical systems and bifurcation theory
M I Camacho, M J Pacifico and F Takens
161 Branched coverings and algebraic functions
M Namba
162 Perturbation bounds for matrix eigenvalues
R Bhatia
163 Defect minimization in operator equations: theory and applications
R Reemtsen
164 Multidimensional Brownian excursions and potential theory
K Burdzy
165 Viscosity solutions and optimal control
R J Elliott
166 Nonlinear partial differential equations and their applications. Collège de France Seminar. Volume VIII
H Brezis and J L Lions
167 Theory and applications of inverse problems
H Haario
168 Energy stability and convection
G P Galdi and B Straughan
169 Additive groups of rings. Volume II
S Feigelstock
170 Numerical analysis 1987
D F Griffiths and G A Watson
171 Surveys of some recent results in operator theory. Volume I
J B Conway and B B Morrel
172 Amenable Banach algebras
J-P Pier
173 Pseudo-orbits of contact forms
A Bahri
174 Poisson algebras and Poisson manifolds
K H Bhaskara and K Viswanath
175 Maximum principles and eigenvalue problems in partial differential equations
P W Schaefer
176 Mathematical analysis of nonlinear, dynamic processes
K U Grusa
177 Cordes' two-parameter spectral representation theory
D F McGhee and R H Picard
178 Equivariant K-theory for proper actions
N C Phillips
179 Elliptic operators, topology and asymptotic methods
J Roe
180 Nonlinear evolution equations
J K Engelbrecht, V E Fridman and E N Pelinovski
181 Nonlinear partial differential equations and their applications. Collège de France Seminar. Volume IX
H Brezis and J L Lions
182 Critical points at infinity in some variational problems
A Bahri
183 Recent developments in hyperbolic equations
L Cattabriga, F Colombini, M K V Murthy and S Spagnolo
184 Optimization and identification of systems governed by evolution equations on Banach space
N U Ahmed
185 Free boundary problems: theory and applications. Volume I
K H Hoffmann and J Sprekels
186 Free boundary problems: theory and applications. Volume II
K H Hoffmann and J Sprekels
187 An introduction to intersection homology theory
F Kirwan
188 Derivatives, nuclei and dimensions on the frame of torsion theories
J S Golan and H Simmons
189 Theory of reproducing kernels and its applications
S Saitoh
190 Volterra integrodifferential equations in Banach spaces and applications
G Da Prato and M Iannelli
191 Nest algebras
K R Davidson
192 Surveys of some recent results in operator theory. Volume II
J B Conway and B B Morrel
193 Nonlinear variational problems. Volume II
A Marino and M K Murthy
194 Stochastic processes with multidimensional parameter
M E Dozzi
195 Prestressed bodies
D Iesan
196 Hilbert space approach to some classical transforms
R H Picard
197 Stochastic calculus in application
J R Norris
198 Radical theory
B J Gardner
199 The C* – algebras of a class of solvable Lie groups
X Wang

200 Stochastic analysis, path integration and dynamics
D Elworthy
201 Riemannian geometry and holonomy groups
S Salamon
202 Strong asymptotics for extremal errors and polynomials associated with Erdös type weights
D S Lubinsky
203 Optimal control of diffusion processes
V S Borkar
204 Rings, modules and radicals
B J Gardner
205 Two-parameter eigenvalue problems in ordinary differential equations
M Faierman
206 Distributions and analytic functions
R D Carmichael and D Mitrović
207 Semicontinuity, relaxation and integral representation in the calculus of variations
G Buttazzo
208 Recent advances in nonlinear elliptic and parabolic problems
P Bénilan, M Chipot, L Evans and M Pierre
209 Model completions, ring representations and the topology of the Pierce sheaf
A Carson
210 Retarded dynamical systems
G Stepan
211 Function spaces, differential operators and nonlinear analysis
L Paivarinta
212 Analytic function theory of one complex variable
C C Yang, Y Komatu and K Niino
213 Elements of stability of visco-elastic fluids
J Dunwoody
214 Jordan decompositions of generalised vector measures
K D Schmidt
215 A mathematical analysis of bending of plates with transverse shear deformation
C Constanda
216 Ordinary and partial differential equations Vol II
B D Sleeman and R J Jarvis
217 Hilbert modules over function algebras
R G Douglas and V I Paulsen
218 Graph colourings
R Wilson and R Nelson
219 Hardy-type inequalities
A Kufner and B Opic
220 Nonlinear partial differential equations and their applications. College de France Seminar Volume X
H Brezis and J L Lions
221 Workshop on dynamical systems
E Shiels and Z Coelho
222 Geometry and analysis in nonlinear dynamics
H W Broer and F Takens
223 Fluid dynamical aspects of combustion theory
M Onofri and A Tesei
224 Approximation of Hilbert space operators. Volume I. 2nd edition
D Herrero
225 Operator Theory: Proceedings of the 1988 GPOTS–Wabash conference
J B Conway and B B Morrel
226 Local cohomology and localization
J L Bueso Montero, B Torrecillas Jover and A Verschoren
227 Nonlinear waves and dissipative effects
D Fusco and A Jeffrey
228 Numerical analysis. Volume III
D F Griffiths and G A Watson
229 Recent developments in structured continua. Volume III
D De Kee and P Kaloni
230 Boolean methods in interpolation and approximation
F J Delvos and W Schempp
231 Further advances in twistor theory, Volume 1
L J Mason and L P Hughston
232 Further advances in twistor theory, Volume 2
L J Mason and L P Hughston
233 Geometry in the neighborhood of invariant manifolds of maps and flows and linearization
U Kirchgraber and K Palmer
234 Quantales and their applications
K I Rosenthal
235 Integral equations and inverse problems
V Petkov and R Lazarov
236 Pseudo-differential operators
S R Simanca
237 A functional analytic approach to statistical experiments
I M Bomze
238 Quantum mechanics, algebras and distributions
D Dubin and M Hennings
239 Hamilton flows and evolution semigroups
J Gzyl
240 Topics in controlled Markov chains
V S Borkar
241 Invariant manifold theory for hydrodynamic transition
S Sritharan
242 Lectures on the spectrum of $L^2(\Gamma\backslash G)$
F L Williams

Harmonic maps into homogeneous spaces

Malcolm Black

University of Warwick

Harmonic maps into homogeneous spaces

Longman
Scientific &
Technical

Copublished in the United States with
John Wiley & Sons, Inc., New York

Longman Scientific & Technical,
Longman Group UK Limited,
Longman House, Burnt Mill, Harlow,
Essex CM20 2JE, England
and Associated Companies throughout the world.

Copublished in the United States with
John Wiley & Sons, Inc., 605 Third Avenue, New York, NY 10158

First published 1991

AMS Subject Classification: (Main) 58E20
(Subsidiary) 53C30

ISSN 0269-3674

British Library Cataloguing in Publication Data
Black, M.
Harmonic maps into homogeneous spaces.
– (Pitman research notes in mathematics)
I. Title II. Series
515

ISBN 0-582-08765-1

Library of Congress Cataloging-in-Publication Data
Black, M. (Malcolm)
Harmonic maps into homogeneous spaces / M. Black.
p. cm. -- (Pitman research notes in mathematics series, ISSN 0269-3674 ; 255)
Includes bibliographical references.
1. Harmonic maps. 2. Homogeneous spaces. I. Title. II. Series.
QA614.73.B53 1991
514'.74--dc20

91-18062
CIP

Printed and bound in Great Britain
by Biddles Ltd, Guildford and King's Lynn

Contents

Chapter 1
Introduction

These notes concern themselves with harmonic maps into homogeneous spaces and in particular concentrate upon maps of Riemann surfaces into flag manifolds and are a slightly revised and extended version of the author's Ph.D. thesis. The contents are briefly summarized in section 1.5.

§1. Harmonic Maps

Let (M, g) and (N, h) be Riemannian manifolds with M compact. A smooth map $\varphi : M \to N$ is said to be *harmonic* if it is a critical point of the energy functional:

$$E(\varphi) = \tfrac{1}{2}\int_M |d\varphi|^2 \nu_g, \tag{1.1}$$

where $|d\varphi|$ denotes the norm of the differential $d\varphi_x \in T_x{}^*M \otimes T_{\varphi(x)}N$ inherited from the Riemannian structures and ν_g is the volume form on M.

The Euler-Lagrange equations for this variational problem are

$$\mathrm{tr}\nabla d\varphi = 0, \tag{1.2}$$

where ∇ is the connection on $TM \otimes \varphi^{-1}TN$ induced by the Levi-Civita connections on TM and TN.

Harmonic maps provide a generalisation of some classical variational problems: those with one dimensional domain are precisely geodesics parametrized by arc length; harmonic maps with co-domain $\mathbb{R}$ are harmonic functions, while a weakly conformal harmonic map of a surface is a branched minimal immersion.

The surveys of Eells and Lemaire [EL1, EL2, EL3] provide an excellent overview of harmonic map theory.

§2. Two Dimensional Domains

The theory of harmonic maps of two dimensional domains is highly non-trivial, but nevertheless has several special features which lead to an especially rich theory. In these notes we shall concentrate (although not exclusively) on harmonic maps of Riemann surfaces. We now outline some of the special features mentioned.

Assume that M is an orientable surface. In the notation of equation (1.1), since M is two dimensional the energy density $|d\varphi|^2 \nu_g$ is invariant under a conformal change of metric. On an orientable surface the choice of a conformal class of metrics is equivalent to determining a complex structure. Hence the property of a map being harmonic depends only on the complex structure induced by the metric - in other words only on the structure of M as a Riemann surface. Choose a chart (U,z) on M compatible with the Riemann surface structure, then the Euler-Lagrange equations (1.2) may be reformulated over this chart as

$$\nabla_{\bar{z}}\, \varphi_*(\frac{\partial}{\partial z}) = 0. \qquad (1.3)$$

Equation (1.3) shows that φ is harmonic precisely when $d\varphi^{1,0}$ is a holomorphic section over M.

In the light of the discussion above it is entirely natural that complex analytic techniques should play such a crucial role in the study of harmonic maps of surfaces. Indeed, holomorphic maps between Kähler manifolds (and Riemann surfaces are always Kähler) are energy minimizing in their homotopy class and therefore harmonic [ES]. Lichnerowicz [L] gave conditions for harmonicity of holomorphic maps of Hermitian manifolds with almost complex structures. This result was extended by Rawnsley [R2] to encompass co-domains with f-structures and f-holomorphic maps.

The heat flow for surfaces is especially well behaved [St]. Although not in general completely regular, the singularities are limited to the bubbling off of a finite number of harmonic spheres. In contrast, the heat flow on three dimensional domains can blow up in finite time [D].

§ 3. Maps into Symmetric Spaces

The study of harmonic maps into symmetric spaces is of particular interest because it is often possible to obtain a concrete description of the harmonic maps in terms of holomorphic data. A further motivation comes from physics in that such harmonic maps are the classical solutions to non-linear σ models. These models are an attempt to construct a more tractable field theory which contains many of the features of the full non-abelian gauge field theories.

The literature concerning this problem is now very large and a recurrent theme in this work is the *twistor method*: harmonic maps of surfaces are realized as the images under a projection of holomorphic maps into a space fibering over the co-domain. A classical twistor result is that a harmonic function on a two-dimensional domain is the real part of a holomorphic function, another is the Weierstrass preparation theorem for minimal surfaces in $\mathbb{R}^3$ [La].

Calabi [C] studied minimal immersions (equivalently, harmonic immersions) of S^2 into spheres. This work provided a novel combination of homogeneous geometry and twistor methods. To each minimal immersion in S^{2n} he associated a holomorphic map of S^2 into $\dfrac{SO(2n+1)}{U(n)}$ and was then able to apply complex analytic methods to achieve a description of these maps. He was able to show that linearly full minimal immersions occur only in even dimensional spheres and that their energy is quantized: ie. it is always an integer multiple of $2\pi r^2$, where r is the radius of the co-domain.

Eells and Wood [EW] elaborated on these ideas to produce a complete description of harmonic maps of S^2 into $\mathbb{C}P^n$. The methods used by Calabi, Eells and Wood have many similarities and were presented in unified form by Burstall and Rawnsley [BR2].

To give a flavour of these ideas, we shall digress briefly to describe the work of [EW] in more detail.

Let φ be a holomorphic map of Riemann surface M into $\mathbb{C}P^n$, and we insist further that φ is linearly full: ie. that φ does not factor through any smaller complex projective space lying in $\mathbb{C}P^n$. $\mathbb{C}P^n$ may be considered as the complex lines in $\mathbb{C}^{n+1}$ and hence there is a covering $\mathbb{C}^{n+1}\backslash\{0\} \rightarrow \mathbb{C}P^n$. Over a sufficiently small chart (U,z) in M we can lift φ to a map $\tilde{\varphi}$ into $\mathbb{C}^{n+1}\backslash\{0\}$. Consider the derivative $\dfrac{\partial\tilde{\varphi}}{\partial z}$ and take the component perpendicular to $\tilde{\varphi}$ to define a new map $\partial\varphi$ of (U,z) into $\mathbb{C}P^n$. Because φ is holomorphic (hence harmonic), it has holomorphic second fundamental form (this is essentially equation (1.3)) and the definition of $\partial\varphi$ can be extended over zeros in the rank of φ. Furthermore, $\partial\varphi$ is independent of the choice of chart and lift and so may be defined globally over M.

It is at this point that we use the twistor technique. The map $\psi = (\varphi,\partial\varphi,(\varphi\oplus\partial\varphi)^{\perp})$ may be regarded as a map into $\dfrac{SU(n+1)}{S(U(1)\times U(1)\times U(n-1))}$. The latter is a flag manifold which fibres naturally over $\mathbb{C}P^n = \dfrac{SU(n+1)}{S(U(1)\times U(n))}$ in two ways: namely by projection onto the first or second factor $(\pi_1, \pi_2$ respectively). The map ψ is both holomorphic and 'horizontal' -its derivatives are constrained to lie in certain horizontal distributions on $\frac{SU(n+1)}{S(U(1)\times U(1)\times U(n-1))}$. The twistor result is that $\pi_2 \circ \psi = \partial\varphi$ is also a harmonic map.

Here then the twistor method produces harmonic maps from holomorphic ones. In fact the method may be extended be applied to harmonic initial data φ. We can therefore perform the process inductively to obtain harmonic maps $\partial\varphi$, $\partial(\partial\varphi) = \partial^2\varphi$ etcetera. The induction stops only when we attempt the procedure on an anti-holomorphic map, in which case the derivative $\frac{\partial\tilde{\varphi}}{\partial z}$ is parallel to $\tilde{\varphi}$.

The main result that Eells and Wood established was that all harmonic maps of S^2 into $\mathbb{CP}^n$ can be produced by this procedure. That is, for each harmonic map ψ there is a holomorphic map φ such that $\psi = \partial^r\varphi$ for some $r \geq 0$.

Much effort was then directed towards understanding harmonic maps of Riemann surfaces into more general symmetric spaces, especially complex Grassmanians[1]. Despite the increased complexity of the situation, several descriptions of harmonic maps of S^2 into complex Grassmanians were achieved.

Complex Grassmanian manifolds admit totally geodesic embeddings in U(n). Uhlenbeck [U] proved that harmonic maps of S^2 into complex Grassmanians factorize (in U(n)) as a product of holomorphic maps into Grassmanians. Valli [V] connected this factorization to a quantized energy reduction at each factor. Burstall and Rawnsley [BR1] were able to extend this technique to deal with more general Hermitian symmetric spaces.

Wood [W] established a concrete, algorithmic approach to constructing the factorization while Burstall and Salamon [BS] provided a twistor description (but with respect to very non-integrable almost complex structures on the twistor space - in contrast to the twistor results of Calabi and Eells - Wood).

[1] See for example [BW, CW, BS, B1, B2.]

§4. Flag Manifolds

In contrast to symmetric spaces, the study of harmonic maps into flag manifolds or more general homogeneous spaces has received relatively little attention. These notes are mainly concerned with this theme.

We have seen that flag manifolds occur as twistor spaces fibring over symmetric spaces. Thus, in one sense our topic is an extension of the work on harmonic maps of Riemann surfaces into symmetric spaces. Indeed, our wider viewpoint will allow us to construct a unified framework for many of the twistor results appearing there. Our subject of study is also of interest in its own right. It has recently been shown that the Higgs field boundary values of the Yang-Mills-Higgs equations on the sphere at infinity provide a harmonic map into a flag manifold [IM].

We now summarize what is known about harmonic maps into flag manifolds. Guest [Gu] investigated maps between flag manifolds induced by group homomorphisms and obtained a Lie algebraic characterization of harmonicity. He produced examples of harmonic maps from full flag manifolds into $\mathbb{C}P^n$ which are harmonic for all the invariant metrics on the full flag - an interesting contrast to the equi-harmonic maps we study here.

Negreiros [N1] considered 'Eells-Wood' maps (i.e. the holomorphic curves in $\dfrac{SU(n)}{S(U(1)^n)}$ produced in [EW] which fibre twistorially over harmonic maps into $\mathbb{C}P^{n-1}$). He established that these maps are harmonic for all the invariant metrics and then calculated for which metrics these maps are stable. Negreiros also produced harmonic, but not almost holomorphic, maps of T^2 into $\dfrac{SU(n)}{S(U(1)^n)}$ with its Killing metric. These maps are equivariant with respect to an S^1 action.

The method of calibrations [HL] was used by Le Khong Van [LKV] to prove that certain homogeneous submanifolds are globally minimal submanifolds in the full flag manifolds of the classical Lie groups.

§5. Summary of Contents

Chapter 2 collects the results on homogeneous geometry we shall need. In particular we identify a class of reductive homogeneous spaces - those whose tangent space splits into distinct isotropy irreducible subspaces - for which a version of Schur's Lemma (Lemma 2.5) holds. We then calculate the invariant metrics, momentum maps and Levi-Civita connections on these spaces. Finally, Section 2.6 sets out the results and ideas we require from the structure theory of compact Lie groups.

Chapter 3 introduces f-structures and establishes conditions for f-holomorphic maps to be harmonic (Theorem 3.1).

In Chapter 4 we investigate invariant f-structures on homogeneous spaces and the conditions under which these are suitable for applications of Theorem 3.1. We say that an f-structure is *horizontal* if it satisfies

$$[F_+, F_-] \subset \mathbf{h}$$

where $F_\pm$ is the $\pm i$ eigenspace and $\mathbf{h}$ is the isotropy algebra. The significance of the definition is that maps which are f-holomorphic with respect to a horizontal f-structure are harmonic for all invariant metrics (Cor. 4.4) when the homogeneous space has distinct irreducible summands. The chapter concludes by illustrating the condition imposed by Theorem 3.1 on f-structures on $SU(n)$ flag manifolds.

One problem that presents itself on considering harmonic maps into homogeneous spaces is to choose the most appropriate metric. The reductive metric (induced by restricting the

Killing form) has the largest isometry group, but is not well behaved from the point of view of complex geometry. On flag manifolds for example, except in degenerate cases, the reductive metric is neither Kähler nor even (1,2) symplectic whatever the choice of almost complex structure. Any other choice of metric involves a reduction of the symmetry of the problem. However, the results of Chapter 4 suggest an alternative approach - namely, to consider maps which are harmonic for all the metrics on the homogeneous space (*equi-harmonic* maps).

Chapter 5 establishes a few basic properties of equi-harmonic maps. Theorem 5.2 derives the Euler-Lagrange equations for equi-harmonic maps from Nöther's Theorem, we also see that equi-harmonicity is preserved by homogeneous diffeomorphisms (Theorem 5.1) and by homogeneous projections (Theorem 5.3). No direct analogue of these results would be available if we chose to study maps harmonic with respect to a single metric. For example, homogeneous diffeomorphisms are isometries for the reductive metric and hence the analogue of Theorem 5.1 holds for reductive harmonic maps. On the other hand reductive harmonic maps are poorly behaved under homogeneous projection.

Chapter 6 addresses the construction and classification of horizontal f-structures on flag manifolds. Firstly (§6.2) we make precise the definition of an irreducible f-structure in such a way as to ensure that the underlying flag geometry is compatible with the decomposition of an f-structure into irreducible pieces. Irreducible horizontal f-structures fall into two categories, the first of which may be classified by elementary root space theory (§6.3). To the remaining class of f-structures we associate a Lie algebra automorphism (§6.4), Kac's theory of finite order Lie algebra automorphisms can then be used to complete our classification. Theorem 6.13 summarizes this classification, expressing an arbitrary horizontal f-structure as a direct sum of irreducible f-structures.

Having identified the horizontal f-structures we can now construct many twistor spaces over homogeneous spaces. Our results unite several previously disparate examples that occur in the literature and provide new examples.

Chapter 7 is devoted to describing horizontal, f-holomorphic group orbits in flag manfolds. First we show that any such orbit is a Hermitian symmetric space and then apply Theorem 6.13 to allow us to find the examples of such orbits in full flag manifolds.

Chapter 8 concerns itself with equi-harmonic maps of Riemann surfaces into full flag manifolds. Provided that such a map be equi-weakly conformal (i.e. weakly conformal for each invariant metric on the co-domain), we obtain a converse to Corollary 4.4 : any equi-minimal (i.e. equi-harmonic and equi-weakly conformal) map of a Riemann surface into a full flag manifold is f-holomorphic with respect to a horizontal f-structure. Theorem 6.13 may then be applied to show that any such equi-minimal map is a direct sum of full horizontal curves (§8.2). The section concludes with a discussion of horizontal curves.

Acknowledgements

I should like to thank my thesis supervisor, John Rawnsley, for suggesting this problem to me and for stimulating conversations. I hope also that John's influence as to how to write mathematics is apparent in these notes. Thanks also to Terri Moss for the efficient typing. Final thanks to all those at Warwick who helped make it such a lively place to work.

Chapter 2
Homogeneous Geometry

§1. Generalities

The approach to homogeneous geometry set out in this chapter follows that of [BR2]. This section sets out the elementary notation and results of this approach and is adapted from the first chapter of [BR2].

Let G be a Lie group and H a closed Lie subgroup. The set of right H cosets $\{gH : g \in G\}$ can be given a manifold structure [Wa]. The *homogeneous space* $G/_H$ consists of this manifold together with the transitive left G action. There is a natural principal H bundle over $G/_H$ with total space G and such that H acts on G by right multiplication. The projection map onto $G/_H$ is simply the quotient map $\pi : G \to G/_H$.

The isotropy group at $\pi(e)$ is H. If W is an H representation space then the associated bundle $G \times_H W$ will be denoted $\underline{W}$.

The infinitesimal version of the G action on $G/_H$ provides the link between Lie algebra theory and the geometry of homogeneous spaces. Specifically, there is a map

$$\alpha : G/_H \times \mathfrak{g} \longrightarrow T\, G/_H,$$

defined by

$$\alpha(p, \xi) = \left.\frac{d}{dt}\right|_{t=0} \exp t\,\xi \cdot p.$$

The map α is a surjective map of vector bundles with kernel $\underline{\mathbf{h}}$, the *isotropy bundle*. The fibre of $\underline{\mathbf{h}}$ at p is $\mathbf{h}_p$, the Lie algebra of the isotropy group at p which we denote by H_p.

A homogeneous space is said to be *reductive* if there is a complement $\underline{\mathbf{m}}$ to $\underline{\mathbf{h}}$ in $G/_H \times \mathbf{g}$ such that for all $p \in G/_H$, $\mathbf{m}_p$ is H_p invariant. Suppose

$$\mathbf{g} = \mathbf{h}_p + \mathbf{m}_p$$

is a direct sum and $\mathbf{m}_p$ is H_p invariant, then conjugation by g provides a suitable reductive complement at $g{\cdot}p$, hence it is sufficient to check the reduction condition over a point. When $G/_H$ is reductive, α can be viewed as an isomorphism of $\underline{\mathbf{m}}$ with $T\,G/_H$, in which case let β be the unique bundle map satisfying:

$$\beta : T\,G/_H \to G/_H \times \mathbf{g},$$
$$\beta \circ \alpha = P_{\underline{\mathbf{m}}} \text{ (the projection onto } \underline{\mathbf{m}} \text{ along } \underline{\mathbf{h}}),$$
$$\alpha \circ \beta = \mathrm{id}_{T\,G/_H}.$$

Viewed as a 1-form on $G/_H$ with values in $\mathbf{g}$, β is known as the *Maurer-Cartan form* of $G/_H$ [BR2].

Let $\underline{\mathbf{g}} = \underline{\mathbf{h}} \oplus \underline{\mathbf{m}}$ be reductive summands, and let $P_{\underline{\mathbf{h}}}$ denote the projection onto $\underline{\mathbf{h}}$ along $\underline{\mathbf{m}}$. The 1-form $P_{\underline{\mathbf{h}}}\theta$, where θ is the left Maurer-Cartan form of G, is a connection form on the principal bundle $\pi : G \to G/_H$. This connection is called the *canonical connection*. Since

$$T\,G/_H \cong \underline{\mathbf{m}} = G \times_H \mathbf{m}_e,$$

the canonical connection induces a covariant differentiation on $T\,G/_H$. Invariant tensors on $G/_H$ correspond to constant sections of the bundle $\underline{\mathbf{m}}$ and are thus parallel with respect to the canonical connection.

Suppose that V is an H–space and the representation of H on V is the restriction of a representation of G. Then $\underline{V}$ may be identified with the trivial bundle ${}^{G}/_{H} \times V$ by the map

$$[\, g, v\,] \mapsto (\pi(g), g\cdot v).$$

In this case, there is a relationship between flat differentation and the covariant differentation ∇^c induced by the canonical connection.

Lemma 2.1 [BR2]. Let $f : {}^{G}/_{H} \to \underline{V}$ be a smooth section of $\underline{V}$, then

$$df = \nabla^c f + \beta\cdot f.$$

Suppose that $V = W_1 + W_2$ is a direct sum with H invariant summands. The splitting is H invariant, and hence parallel with respect to ∇^c.

Lemma 2.2 [BR2]. Denote the representation of G on V by $\sigma : G \to \mathrm{End}(V)$, and let $\underline{V} = \underline{W}_1 \oplus \underline{W}_2$. Let $P_i : \underline{V} \to \underline{W}_i$ be the projection onto $\underline{W}_i$ viewed as a function $P_i : {}^{G}/_{H} \to \mathrm{End}(V)$. Then

$$dP_i = [\, \sigma(\beta), P_i\,].$$

The analogue of the structure equations for β is given by the next lemma.

Lemma 2.3 [BR2]

$$d\beta = (1 - \tfrac{1}{2} P_{\mathfrak{m}})[\, \beta \wedge \beta\,], \qquad (2.1)$$

$$\beta \circ T^c = -\tfrac{1}{2} P_{\mathfrak{m}} [\, \beta \wedge \beta\,], \qquad (2.2)$$

where T^c is the torsion of the canonical connection on $\underline{\mathfrak{m}}$.

§2. Reductive Splittings

Over a reductive homogeneous space there is a decomposition of $\mathbf{g}$:

$$\underline{\mathbf{g}} = \underline{\mathbf{h}} \oplus \underline{\mathbf{m}} ;$$

such that each of the fibres $\mathbf{h}_x$ and $\mathbf{m}_x$ are H_x invariant.

An H-space is said to be *completely reducible* [J] if it splits as a direct sum of irreducible H spaces. We call a choice of such summands a *reductive splitting*.

Definition. A homogeneous space is *completely reductive* if $\underline{\mathbf{m}}$ is a completely reducible H_x space at each point x.

Suppose we can choose a reductive splitting at a point x; then applying the automorphism Ad_g of $\mathbf{g}$ provides a reductive splitting at $\mathbf{m}_{g \cdot x}$. Hence it is sufficient to test complete reductiveness over a point in G/H to obtain a global reductive splitting:

$$\underline{\mathbf{g}} = \underline{\mathbf{h}} \oplus \underline{\mathbf{m}} , \; \underline{\mathbf{m}} = \bigoplus_{\alpha \in A} \underline{\mathbf{m}}_\alpha ; \qquad (2.3)$$

with each $\mathbf{m}_{\alpha,x}$ an irreducible H_x space.

We call a group H *strongly reductive* if every H representation is completely reducible. This condition is slightly stronger than the more usual condition 'reductive', which is that every representation of the Lie algebra of H is completely reducible. Note that both connected reductive Lie groups and compact Lie groups are strongly reductive.

§3. Distinct Reductive Summands

It will sometimes be necessary to restrict our attention to homogeneous spaces such that the irreducible summands $\mathbf{m}_\alpha$ are mutually distinct H spaces. Large classes of homogeneous spaces satisfy this condition, some examples of which are given below.

(i) Let G be a compact, semi-simple Lie group and T a maximal torus. Elementary root space theory (§2.6) shows that G/T has distinct irreducible summands.

(ii) $\dfrac{SO(n)}{S(\mathbb{Z}_2^n)}$. Here $S(\mathbb{Z}_2^n)$ is the subgroup of SO(n) consisting of the diagonal matrices with diagonal entries equal to ± 1.

(iii) Let $(\mathbf{g}, \sigma)$ be an orthogonal symmetric Lie algebra ([He], [KN]) with $\mathbf{g}$ semi-simple. Explicitly:

(a) $\mathbf{g}$ is a semi-simple Lie algebra,

(b) σ is an involutive automorphism of $\mathbf{g}$,

(c) $\mathbf{h}$ is the fixed point set of σ (and hence a subalgebra of $\mathbf{g}$),

(d) $AD_G(\mathbf{h})$ is compact.

Then we may apply the results[1] in [KN] to prove that the Riemannian globally symmetric space G/H has distinct irreducible summands.

Lemma 2.4. Let H_1 and H_2 be strongly reductive subgroups of G and H_1 a subgroup of H_2. If G/H_1 has distinct irreducible summands then G/H_2 also has distinct irreducible summands.

---

[1] Vol II, Chap. XI, 5.2, 7.3, 7.4

Proof. Let

$$\mathbf{g} = \mathbf{h}_2 \oplus \sum_{\alpha \in A} \mathbf{m}_\alpha$$

be a reductive splitting for $G/_{H_2}$.

$\mathbf{m}_\alpha$ is H_2 invariant, hence H_1 invariant. It follows that there is an H_1 irreducible splitting of $\mathbf{m}_2$:

$$\mathbf{m}_2 = \sum_{\alpha \in A} \sum_{\beta \in B(\alpha)} \mathbf{m}_{\alpha\beta}; \quad \mathbf{m}_\alpha = \sum_{\beta \in B(\alpha)} \mathbf{m}_{\alpha\beta}.$$

The hypothesis on $G/_{H_1}$ implies that the $\mathbf{m}_{\alpha\beta}$ are mutually distinct H_1 spaces. Hence the $\mathbf{m}_\alpha$ are mutually distinct H_1 spaces and thus mutually distinct H_2 spaces. □

Let G be a semi simple Lie group. The flag manifolds of G are the homogeneous spaces $G/_{C(S)}$, where C(S) is the centralizer of the subtorus S of G. C(S) contains a maximal torus of G and so we can apply Lemma 2.4 and example (i) to show that the flag manifolds of G have mutually distinct irreducible summands.

The significance of distinct irreducible summands is that a form of Schur's Lemma holds.

Lemma 2.5. Let $\mathbf{m}$ be a completely reducible H space over a field k whose irreducible summands are mutually distinct H spaces. Index the summands by a set A. Let φ be an H equivariant endomorphism of $\mathbf{m}$ with eigenvalues lying in k. Then

$$\varphi = \sum_{\alpha \in A} \varphi_\alpha P_\alpha, \quad \varphi_\alpha \in k \quad \forall \alpha \in A;$$

where P_α is the projection onto the irreducible summand $\mathbf{m}_\alpha$ of $\mathbf{m}$.

Proof. Use the fact that each P_α is H equivariant and apply Schur's Lemma to $P_\alpha \circ \varphi \circ P_\beta$.

Remark. One immediate consequence of the Lemma is that the reductive splitting of such an **m** is unique.

§4. Invariant Tensors

Let G/H be a completely reductive homogeneous space and choose $x \in G/H$. Restricting a G invariant tensor on G/H to x provides an H_x invariant tensor on $\mathbf{m}_x$. Conversely, H_x invariant tensors on $\mathbf{m}_x$ can be extended by equivariance to G invariant tensors on G/H. This correspondence is a bijection, and hereafter will be used without comment.

Fix a reductive splitting as in equation (2.3). The most basic examples of invariant tensors are the projection maps P_α. Define

$$\beta_\alpha = P_\alpha \circ \beta,$$

where β is the Maurer-Cartan form of G/H (§2.1).

Lemma 2.6. $$d\beta_\alpha = [\beta \wedge \beta_\alpha] - \tfrac{1}{2} P_\alpha[\beta \wedge \beta].$$

Proof.
$$\begin{aligned} d\beta_\alpha &= d(P_\alpha \circ \beta) \\ &= dP_\alpha \wedge \beta + P_\alpha\, d\beta. \end{aligned}$$

Now use Lemmas 2.2 and 2.3 to see that

$$\begin{aligned} d\beta_\alpha &= [\beta \wedge P_\alpha\beta] - P_\alpha[\beta \wedge \beta] + P_\alpha (1 - \tfrac{1}{2} P_{\mathbf{m}}) [\beta \wedge \beta] \\ &= [\beta \wedge P_\alpha\beta] - \tfrac{1}{2} P_\alpha [\beta \wedge \beta]. \end{aligned}$$ □

Note. This lemma also holds for the irreducible splitting of $T^{\mathbb{C}}\, {}^{G}/_{H}$.

For the remainder of this section we shall assume G carries a bi-invariant metric denoted by $\langle\, ,\, \rangle$. Further suppose that ${}^{G}/H$ has distinct irreducible summands. We shall derive formulae for the invariant metrics and momentum maps of ${}^{G}/_{H}$.

Let λ be an invariant metric on ${}^{G}/_{H}$.

Define

$$\tilde{\lambda} : \mathbf{m} \to \mathbf{m}$$

by

$$\langle \tilde{\lambda}(\xi), \eta \rangle = \lambda(\xi, \eta).$$

It follows that $\tilde{\lambda}$ is H equivariant with strictly positive, real eigenvalues. Apply Lemma 2.5 to see that

$$\tilde{\lambda} = \sum_{\alpha \in A} \lambda_\alpha P_\alpha, \quad \lambda_\alpha > 0. \tag{2.4}$$

The *momentum map* of $({}^{G}/_{H}, \lambda)$ is defined as the 1-form μ with values in $\mathbf{g}$ characterized by:

$$\langle \mu(X), \xi \rangle = \lambda(X, \tilde{\xi}), \ X \in \Gamma(T\, {}^{G}/_{H}), \ \xi \in \mathbf{g}. \tag{2.5}$$

For $\xi \in \mathbf{g}$, the vector field $\tilde{\xi}$ is defined by $\tilde{\xi}_p = \left.\frac{d}{dt}\right|_{t=0} \exp t\xi \cdot p.$

This characterization implies that

$$L_g^* \mu = \mathrm{Adg} \cdot \mu, \ \forall g \in G. \tag{2.6}$$

When λ is expressed in the form (2.4) with respect to the reductive splitting, then

$$\mu = \sum_{\alpha \in A} \lambda_\alpha P_\alpha \cdot \beta \tag{2.7}$$

satisfies the characterization (2.5).

If we drop the assumption that G/H has distinct irreducible summands then the formula (2.4) still defines an invariant metric, and (2.7) defines its momentum map. However, not all invariant metrics can be expressed this way with respect to a single reductive splitting.

§5. The Levi-Civita Connection

In this section we derive a formula for the Levi-Civita connection of an invariant metric λ expressed in the form (2.4) with respect to a reductive splitting (2.3).

Let X, Y, Z be vector fields on G/H. In the discussion that follows we use the isomorphism β to alternate between working on $T\, G/H$ and $\mathbf{m}$. We will simplify formulae by suppressing explicit indication of the use of this isomorphism. Note however that Lie brackets are used exclusively to denote the Lie bracket in $\mathbf{g}$, never vector field Lie bracket. $P_{\mathbf{m}}[X, Y]$ will be abbreviated by $[X, Y]_{\mathbf{m}}$.

The canonical connection was defined in §2.1. To define any other connection ∇ on G/H it is sufficient to determine the tensor

$$D(X, Y) = \nabla_X Y - \nabla^c_X Y. \tag{2.8}$$

Observe that ∇ is a G invariant connection iff D is a G invariant tensor.

It is well known that[2] if λ is an invariant (and possibly indefinite) metric on a reductive homogeneous space then:

[2] See for example [KN] Vol. II, Chap. X, theorem 3.3.

$$2\lambda(D(X, Y) - \tfrac{1}{2}[X,Y]_{\mathbf{m}}, Z) = \lambda(X, [Z, Y]_{\mathbf{m}}) + \lambda([Z, X]_{\mathbf{m}}, Y). \tag{2.9}$$

Now suppose $X \in \Gamma(\mathbf{m}_\alpha)$, $Y \in \Gamma(\mathbf{m}_\beta)$ and $Z \in \Gamma(\mathbf{m}_\gamma)$. Apply the characterization (2.4) of the metric λ to (2.9) to obtain:

$$2\lambda_\gamma\langle D(X, Y) - \tfrac{1}{2}[X, Y], Z\rangle = \lambda_\alpha\langle X, [Z, Y]\rangle + \lambda_\beta\langle [Z, X], Y\rangle. \tag{2.10}$$

Now use the Ad invariance of $\langle\, , \rangle$ to see

$$\langle D(X, Y), Z\rangle = \langle \frac{\lambda_\gamma + \lambda_\beta - \lambda_\alpha}{2\lambda_\gamma}[X, Y], Z\rangle, \tag{2.11}$$

and hence that

$$D(X, Y) = \sum_{\gamma \in A} \frac{\lambda_\gamma + \lambda_\beta - \lambda_\alpha}{2\lambda_\gamma} P_\gamma[X, Y] \qquad X \in \mathbf{m}_\alpha,\ Y \in \mathbf{m}_\beta. \tag{2.12}$$

In particular, when λ is the reductive metric

$$D(X, Y) = \tfrac{1}{2}[X, Y]_{\mathbf{m}}. \tag{2.13}$$

Notice that this equation extends complex linearly to $\mathbf{g}^{\mathbb{C}}$, hence is also applicable to complex vectors.

§6. Roots

In the final three chapters we require some Lie group structure theory. The results presented here were collected from [Hu], [He] and [BR2]. The first two references contain a more complete exposition as well as the proofs and standard definitions not included here.

Let $\mathbf{g}^{\mathbb{C}}$ be a semi-simple Lie algebra over $\mathbb{C}$ and choose a Cartan sub-algebra $\mathbf{a}$ (i.e. a maximal set of commuting and semi-simple elements). Denote the Killing form on $\mathbf{g}^{\mathbb{C}}$ by B.

Let α lie in the dual space $\mathbf{a}^*$, and set

$$\mathbf{g}^{\alpha} = \{X \in \mathbf{g}^{\mathbb{C}} : [H, X] = \alpha(H)X \ \forall H \in \mathbf{a}\}.$$

Then $\mathbf{g}^0 = \mathbf{a}$ and the non-zero α with $\mathbf{g}^{\alpha} \neq 0$ are called *roots* with *root spaces* $\mathbf{g}^{\alpha}$. The set of roots is denoted $\Delta(\mathbf{g}^{\mathbb{C}}, \mathbf{a})$.

Theorem 2.7.

(i) $\mathbf{g}^{\mathbb{C}} = \mathbf{a} + \sum_{\alpha \in \Delta(\mathbf{g}^{\mathbb{C}}, \mathbf{a})} \mathbf{g}^{\alpha}$ is a direct sum.

(ii) $\dim \mathbf{g}^{\alpha} = 1 \quad \forall \alpha \in \Delta(\mathbf{g}^{\mathbb{C}}, \mathbf{a})$.

(iii) If $\alpha, \beta \in \Delta(\mathbf{g}^{\mathbb{C}}, \mathbf{a})$ with $\alpha + \beta \neq 0$ then $B(\mathbf{g}^{\alpha}, \mathbf{g}^{\beta}) = 0$.

(iv) B is non-degenerate on $\mathbf{a}$, whence for $\alpha \in \mathbf{a}^*$ there is a unique $H_{\alpha} \in \mathbf{a}$ such that $\alpha(H) = B(H_{\alpha}, H)$ for all $H \in \mathbf{a}$.

Set $\langle\alpha, \beta\rangle = B(H_{\alpha}, H_{\beta})$ for $\alpha, \beta \in \Delta(\mathbf{g}^{\mathbb{C}}, \mathbf{a})$.

(v) If $\alpha \in \Delta(\mathbf{g}^{\mathbb{C}}, \mathbf{a})$ then $-\alpha \in \Delta(\mathbf{g}^{\mathbb{C}}, \mathbf{a})$ and for $X \in \mathbf{g}^{\alpha}$, $Y \in \mathbf{g}^{-\alpha}$ we have $[X, Y] = B(X, Y)H_{\alpha}$.

(vi) Suppose $\alpha, \beta, \alpha + \beta \in \Delta(\mathbf{g}^{\mathbb{C}}, \mathbf{a})$, then

$$[\mathbf{g}^{\alpha}, \mathbf{g}^{\beta}] = \mathbf{g}^{a+\beta}.$$

Let $\alpha \in \Delta(g^{\mathbb{C}}, a)$ and β be any root or zero. The α*-series containing* β is by definition the set of all roots of the form $\beta + n\alpha$, $n \in \mathbb{Z}$.

Theorem 2.8. $\alpha \in \Delta(g^{\mathbb{C}}, a)$, $\beta \in \Delta(g^{\mathbb{C}}, a) \cup \{0\}$.

(i) The α-series containing β has the form $\beta + n\alpha$ $(p \leq n \leq q)$ (i.e. the α-series is an uninterrupted string). Furthermore

$$-2\frac{\langle \beta, \alpha \rangle}{\langle \alpha, \alpha \rangle} = p + q.$$

(ii) The only roots proportional to α are $-\alpha, 0, \alpha$.

(iii) The maximum length of root strings is 4, i.e. $q - p \leq 3$.

Lemma 2.9. Let $g^{\mathbb{C}}$ be the complexification of a compact, real Lie algebra g with maximal torus t. Then $t^{\mathbb{C}}$ is a Cartan subalgebra of $g^{\mathbb{C}}$ and we can choose bases $\{X_\alpha : \alpha \in \Delta(g^{\mathbb{C}}, t^{\mathbb{C}})\}$ for the complement of $t^{\mathbb{C}}$ in $g^{\mathbb{C}}$ such that:

(i) $X_\alpha \in g^\alpha$;

(ii) $\overline{X}_\alpha = X_{-\alpha}$;

(iii) $B(X_\alpha, X_{-\alpha}) = 1$; $B(X_\alpha, X_\beta) = 0$ $\alpha + \beta \neq 0$.

Define $N_{\alpha,\beta}$ by

$$[X_\alpha, X_\beta] = N_{\alpha,\beta} X_{\alpha+\beta}, \text{ then}$$

(iv) $\alpha, \beta, \alpha + \beta \in \Delta(g^{\mathbb{C}}, t^{\mathbb{C}}) \Leftrightarrow N_{\alpha,\beta} \neq 0$.

Proof. Follows easily from previous theorems and the fact that since $g^{\mathbb{C}}$ is the complexification of a compact real Lie algebra we have $\overline{g^\alpha} = g^{-\alpha}$.

A subset S of $\Delta(g^{\mathbb{C}}, t^{\mathbb{C}})$ is said to be *closed* is whenever $\alpha, \beta \in S$ and $\alpha + \beta \in \Delta(g^{\mathbb{C}}, t^{\mathbb{C}})$ then $\alpha + \beta \in S$.

Definition A *positive root system* is a subset Δ^+ of $\Delta(g^{\mathbb{C}}, t^{\mathbb{C}})$ such that

(i) $\Delta^+ \cap -\Delta^+ = \emptyset$,

(ii) Δ^+ is closed,

(iii) $\Delta^+ \cup -\Delta^+ = \Delta(g^{\mathbb{C}}, t^{\mathbb{C}})$.

The elements of Δ^+ are called *positive roots*.

Definition. Given a positive root system, a positive root is *simple* if it cannot be written as a non-trivial sum of positive roots.

Lemma 2.10. Let Δ^+ be a positive root system in $\Delta(g^{\mathbb{C}}, t^{\mathbb{C}})$ and $\alpha_1,\ldots,\alpha_\ell$ the simple roots. Then

(i) $\{\alpha_1,\ldots,\alpha_\ell\}$ are a basis for $(t^{\mathbb{C}})^*$ over $\mathbb{C}$;

(ii) if $\alpha \in \Delta^+$ then α may be written

$$\alpha = \alpha_{i_1} + \alpha_{i_2} + \ldots + \alpha_{i_r} \qquad 1 \le i_j \le \ell,$$

$$= \sum_{i=1}^{\ell} n_i \alpha_i ;$$

where in the first equation each partial sum of the form

$$\alpha_{i_1} + \ldots + \alpha_{i_s} \quad 1 \le s \le r$$

is a root and in the second equation each n_i is a non-negative integer.

Simple Lie Algebras

A root system $\Delta(g^{\mathbb{C}}, t^{\mathbb{C}})$ is *irreducible* if it cannot be partitioned into mutually orthogonal, non-empty subsets. $g^{\mathbb{C}}$ is simple precisely when $\Delta(g^{\mathbb{C}}, t^{\mathbb{C}})$ is irreducible.

Lemma 2.11. Let $g^{\mathbb{C}}$ be simple and Δ^+ a positive root system for $\Delta(g^{\mathbb{C}}, t^{\mathbb{C}})$. Define a partial ordering on $\Delta(g^{\mathbb{C}}, t^{\mathbb{C}})$ by setting $\alpha \leq \beta$ iff $\beta - \alpha$ is zero or a positive linear combination of simple roots. Then there is a $\theta \in \Delta^+$ which is the unique maximal element with respect to this ordering. θ is called the *highest root.*

§7. Nöther's Theorem

Rawnsley and Pluzhnikov [R1], [P] observed that the Euler–Lagrange equations for harmonic maps into homogeneous spaces have an especially simple formulation. Let μ be the momentum map of $(G/H, \lambda)$ (§2.4).

Theorem 2.12. Let φ be a map from Riemannian manifold (N, h) to homogeneous space $(G/H, \lambda)$, where λ is an invariant metric. Then φ is harmonic iff

$$d^* \varphi^* \mu = 0. \tag{2.14}$$

Chapter 3
f-Structures and f-Holomorphic Maps

We introduce here an extension of the notion of an almost complex structure on a manifold namely 'f-structure'. f-structures were first considered by Yano [Yano].

Definition. [Yano] Let $F \in \Gamma(\text{End } TM)$ such that $F^3 + F = 0$. The endomorphism F is called an *f-structure* on M.

The definition shows that F has three possible eigenvalues: 0 and $\pm i$. An almost complex structure on a manifold is an f-structure with trivial 0-eigenspace.

The following theorem is a mild extension of a result of Rawnsley [R2] which in turn had its origins in Lichnerowicz' Proposition [L].

Theorem 3.1 Let $\varphi : (M, g, J) \longrightarrow (N, h, F)$ such that

(i) φ is f-holomorphic i.e. $d\varphi \circ J = F \circ d\varphi$

(ii) $d(*\omega) = 0$

(iii) $(d^{\nabla}F)^{(1,1)} = 0$

then φ is harmonic.

Here (M, g, J) is a 2m-dimensional Riemannian manifold with almost complex structure J, and Kähler form ω. (N, h, F) is a Riemannian manifold with f-structure 'F'. d^{∇} denotes the covariant exterior derivative with respect to the Levi-Civita connection ∇ of h. We will also use $\{\, ,\}$ for the metric h.

Proof. The map φ is harmonic $\Leftrightarrow d^{\nabla}{*}d\varphi = 0$ [EL2] where $d\varphi \in \Gamma(T^*M \otimes \varphi^{-1}TN)$.

Claim. $2{*}d\varphi = d\varphi \circ J \wedge {*}\omega$.

Proof of Claim. Let $\alpha \in \Gamma(T^*M \otimes \varphi^{-1}TN)$ then

$$\{\alpha \wedge d\varphi \circ J \wedge {*}\omega\} = \{\alpha \wedge d\varphi \circ J\} \wedge {*}\omega$$

$$= <\{\alpha \wedge d\varphi \circ J\},\ \omega> \mathrm{vol}$$

$$= 2<\alpha, d\varphi>\mathrm{vol}.$$

The last equality may be established by choosing a local orthonormal basis for M adapted to the almost complex structure.

Returning to the main proof we now have:

$$2d^{\nabla}{*}d\varphi = d^{\nabla}(d\varphi \circ J \wedge {*}\omega)$$

$$= d^{\nabla}(d\varphi \circ J) \wedge {*}\omega - d\varphi \circ J \wedge d({*}\omega)$$

$$= d^{\nabla}(d\varphi \circ J) \wedge {*}\omega \qquad \text{from (ii).}$$

${*}\omega$ is an $(m-1, m-1)$ form on M, so to ensure that this last expression vanishes it is sufficient that $(d^{\nabla}(d\varphi \circ J))^{(1,1)} = 0$.

Now, regarding F as a TN valued 1-form, (i) is equivalent to: $d\varphi \circ J = \varphi^*F$. Also $d^{\nabla}(\varphi^*F) = \varphi^*d^{\nabla}F$. Here φ^* and d^{∇} commute since the connection used to define d^{∇} on $\varphi^{-1}TN$ was the pull-back of the Levi-Civita connection on TN.

Applying (i) once more

$$(d^{\nabla}(\varphi^*F))^{1,1} = \varphi^*((d^{\nabla}F)^{1,1}).$$

The condition (iii) now gives the result. □

It will be convenient to characterize the above condition (iii) differently. Let $X \in \Gamma(F_+)$ and $Y \in \Gamma(F_-)$, then

$$\begin{aligned} d^{\nabla}(X,Y) &= \nabla_X F(Y) - \nabla_Y F(X) - F([X,Y]) \\ &= - i\nabla_X Y - i\nabla_Y X - F(\nabla_X Y - \nabla_Y X) \\ &= - (F + i)\nabla_X Y + (F - i)\nabla_Y X. \end{aligned} \tag{3.1}$$

Let P_{F_+} denote projection onto F_+. Note that

$$- 2P_{F_+} = F(F + i),$$

and similarly

$$P_{F_0} = (F + i)(F - i).$$

We see that

$$P_{F_+} d^{\nabla}F(X, Y) = 2iF(F + i)\nabla_X Y \tag{3.2}$$

and

$$P_{F_0} d^{\nabla}F(X, Y) = -i(F^2 + 1)(\nabla_X Y + \nabla_Y X). \tag{3.3}$$

These calculations have now proved:

Lemma 3.2. $(d^{\nabla}F)^{1,1} = 0$ iff

(i) $\nabla_{F_+}\Gamma(F_-) \subset \Gamma(F_0 \oplus F_-)$

and

(ii) $P_{F_0}(\nabla_X Y + \nabla_Y X) = 0$ $\quad\quad \forall X \in \Gamma(F_+)$ and $\forall Y \in \Gamma(F_-)$.

Note. By conjugation (i) $\Leftrightarrow$ (i′) $\nabla_{F_-}\Gamma(F_+) \subset \Gamma(F_0 \oplus F_+)$.

When F is actually an almost complex structure such that $\omega(X, Y) = g(X, FY)$ is anti-symmetric, then the results of Rawnsley and Salamon ([R2], [S]) show that $(d^\nabla F)^{1,1} = 0$ precisely when $d\omega^{1,2} = 0$. In this way Lichnerowicz' proposition may be seen as the case $F_0 = \{0\}$ of Theorem 3.1.

Chapter 4
f-Structures on Reductive Homogeneous Spaces

§1. f-Structures and Metrics

Reductive homogeneous spaces provide a rich variety of interesting f-structures which we start to examine in this section.

Let G/H be a completely reductive homogeneous space with reductive splitting:

$$\mathbf{g} = \mathbf{h} \oplus \mathbf{m}, \quad \mathbf{m} = \bigoplus_{\alpha \in A} \mathbf{m}_\alpha. \tag{4.1}$$

It will be necessary to consider the complexified version of (4.1). It is certainly true that $\mathbf{m}_\alpha^{\mathbb{C}}$ is H invariant. Suppose that $\mathbf{m}_\alpha^1$ is an H invariant subspace of $\mathbf{m}_\alpha^{\mathbb{C}}$, then so are $\mathrm{Re}\,(\mathbf{m}_\alpha^1 + \overline{\mathbf{m}}_\alpha^1)$ and $\mathrm{Re}\,(\mathbf{m}_\alpha^1 \cap \overline{\mathbf{m}}_\alpha^1)$ where $\overline{\mathbf{m}}_\alpha^1$ is the conjugate of $\mathbf{m}_\alpha^1$. We have:

$$\mathrm{Re}\,(\mathbf{m}_\alpha^1 + \overline{\mathbf{m}}_\alpha^1) = \mathbf{m}_\alpha$$

and

$$\mathrm{Re}\,(\mathbf{m}_\alpha^1 \cap \overline{\mathbf{m}}_\alpha^1) = \{0\} \text{ or } \mathbf{m}_\alpha.$$

Thus either

$$\mathbf{m}_\alpha^{\mathbb{C}} \text{ is H irreducible} \tag{4.2}$$

or

$$\mathbf{m}_\alpha^{\mathbb{C}} = \mathbf{m}_\alpha^1 \oplus \mathbf{m}_\alpha^2, \text{ where } \mathbf{m}_\alpha^2 = \overline{\mathbf{m}}_\alpha^1. \tag{4.3}$$

We have shown that:

$$\mathbf{m}^{\mathbb{C}} = \bigoplus_{\alpha \in B} \mathbf{m}_\alpha^{\mathbb{C}} \oplus \bigoplus_{\alpha \in \Gamma} (\mathbf{m}_\alpha^1 \oplus \mathbf{m}_\alpha^2), \tag{4.4}$$

$B = \{\alpha \in A \mid \mathbf{m}_\alpha^{\mathbb{C}} \text{ is } H \text{ irreducible}\}$,

$\Gamma = \{\alpha \in A \mid \mathbf{m}_\alpha^{\mathbb{C}} \text{ not } H \text{ irreducible}\}$.

An invariant f-structure on $G/_H$ may be identified with an H equivariant endomorphism, F, of $\mathbf{m}$ such that $F^3 + F = 0$. Such a map induces a splitting of $\mathbf{m}^{\mathbb{C}}$ into H invarianteigenspaces:

$\mathbf{m}^{\mathbb{C}} = F_+ \oplus F_- \oplus F_0$;

$F_+ = +i$ eigenspace,

$F_- = \overline{F_+} = -i$ eigenspace.

Identifying suitable eigenspaces is equivalent to defining an f-structure.

Invariant f-structures may be constructed by considering (4.4). Define F_+ to be a sum of spaces $\mathbf{m}_\alpha^i$ (where $\alpha \in \Gamma$) such that if $\mathbf{m}_\alpha^1 \subset F_+$ then $\mathbf{m}_\alpha^2 \not\subset F_+$ and vice versa; this ensures that $\overline{F_+} \cap F_+ = \{0\}$. Define $F_- = \overline{F_+}$ and let the remaining irreducible summands of $\mathbf{m}^{\mathbb{C}}$ constitute F_0. Determining the eigenspaces in this way defines an H equivariant endomorphism F of $\mathbf{m}^{\mathbb{C}}$, which is seen to be the $\mathbb{C}$-linear extension of an H equivariant endomorphism of $\mathbf{m}$ since F_+ and F_- are conjugate.

Remark 1. In the case where the $\mathbf{m}_\alpha$ are mutually distinct H spaces, Lemma 2.5 ensures that this construction produces all the invariant f-structures on such homogeneous spaces.

Let us now state the version of Lemma 3.2 appropriate to reductive homogeneous spaces. Let ∇ be the Levi-Civita connection of an invariant metric Λ on a reductive homogeneous space. Set $D = \nabla - \nabla^c$.

Lemma 4.1. $(d^{\nabla}F)^{1,1} = 0$ iff:

(i) $D(F_+, F_-) \subset F_0 \oplus F_-$

and

(ii) $P_{F_0}\{D(X, \bar{Y}) + D(\bar{Y}, X)\} = 0 \ \ \forall X, Y \in F_+$.

Proof. Recall that invariant tensors are parallel for the canonical connection and apply Lemma 3.2. □

Suppose now that we can choose a reductive splitting for **m** such that the metric is of the form of equation 2.4 and the f-structure is of the type described above. In this case we shall say that the metric and f-structure are *compatible*.

Remark 2. When the $\mathbf{m}_\alpha$ are mutually distinct H spaces, Lemma 2.5 ensures that all metrics and f-structures are automatically compatible with respect to the unique reductive splitting.

Lemma 4.2. Suppose that f-structure F, and metric Λ are compatible on $^G/_H$. Then $(\mathbf{h}, \mathbf{m}, F, \Lambda)$ satisfies $(d^\nabla F)^{1,1} = 0$ iff:

whenever $\mathbf{m}^i_\alpha$ is contained in F_+ and $\mathbf{m}^j_\beta$ is contained in F_- then

(i) if $P_{F_0}[\mathbf{m}^i_\alpha, \mathbf{m}^j_\beta] \neq \{0\}$ then $\lambda_\alpha = \lambda_\beta$

and

(ii) if $P^k_\gamma[\mathbf{m}^i_\alpha, \mathbf{m}^j_\beta] \neq \{0\}$ and $\mathbf{m}^k_\gamma \subset F_+$ then $\lambda_\alpha = \lambda_\beta + \lambda_\gamma$.

Proof. Let $X \in \mathbf{m}^i_\alpha$, $Y \in \mathbf{m}^j_\beta$. Equation 2.12 shows that

$$D(X, Y) = \sum_{\mathbf{m}^k_\gamma \subset \mathbf{m}} \frac{\lambda_\gamma + \lambda_\beta - \lambda_\alpha}{2\lambda_\gamma} P^k_\gamma[X, Y].$$

Thus

$$P_{F_+} D(X, Y) = \sum_{\mathbf{m}^k_\gamma \subset F_+} \frac{\lambda_\gamma + \lambda_\beta - \lambda_\alpha}{2\lambda_\gamma} P^k_\gamma[X, Y]; \tag{4.5}$$

and

$$P_{F_0}\{D(X, Y) + D(Y, X)\} \tag{4.6}$$

$$= \sum_{\mathbf{m}^k_\gamma \subset F_0} \left\{ \frac{\lambda_\gamma + \lambda_\beta - \lambda_\alpha}{2\lambda_\gamma} P^k_\gamma[X, Y] + \frac{\lambda_\gamma + \lambda_\alpha - \lambda_\beta}{2\lambda_\gamma} P^k_\gamma[Y, X] \right\}$$

$$= \sum_{\mathbf{m}^k_\gamma \subset F_0} \frac{\lambda_\beta - \lambda_\alpha}{\lambda_\gamma} P^k_\gamma[X, Y].$$

Applying Lemma 4.1 now gives the result. □

§2 : Horizontality

Lemma 4.2 has an immediate and important consequence.

Definition. An invariant f-structure on a reductive homogeneous space with the property that $[F_+, F_-] \subset \mathbf{h}$ will be called a *horizontal* f-*structure.*

Theorem 4.3. Let $\varphi : (M, g, J) \to G/H$ be a map f-holomorphic with respect to a horizontal f-structure and assume $d(*\omega) = 0$ where ω is the Kähler form on M. Then φ is harmonic with respect to any metric compatible with F.

Proof. Apply Lemma 4.2 and Theorem 3.1. □

Corollary 4.4. Suppose the tangent space of G/H has mutually distinct irreducible summands. If φ satisfies the hypotheses of the theorem then φ is harmonic with respect to all the invariant metrices on G/H.

Proof. See Remark 2.

Much of this thesis will concern itself with horizontal f-structures. Here is a first example.

Example

Let σ be an automorphism of Lie algebra $\mathbf{g}$, acting orthogonally with respect to a bi-invariant metric on $\mathbf{g}$. This last condition is automatic if $\mathbf{g}$ is compact and simple.

Let Δ be the set of eigenvalues of σ, and σ^α the eigenspace with eigenvalue α in $\mathbf{g}^{\mathbb{C}}$. Orthogonality ensures that $\Delta \subset S^1$ and $\overline{\sigma^\alpha} = \sigma^{\bar{\alpha}}$. So

$$\mathbf{g}^{\mathbb{C}} = \bigoplus_{\alpha \in \Delta} \sigma^{\alpha},$$

where $\oplus$ indicates orthogonal direct sum with respect to the Hermitian extension of the metric on $\mathbf{g}$ to $\mathbf{g}^{\mathbb{C}}$.

Note also that $[\sigma^{\alpha}, \sigma^{\beta}] \subset \sigma^{\alpha\beta}$ and hence each eigenspace σ^{α} is σ^1 invariant. In particular the real part of σ^1 is a sub-algebra of $\mathbf{g}$.

Set $F_+ = \sigma^{\alpha}$, $F_- = \overline{\sigma^{\alpha}} = \sigma^{\bar{\alpha}}$ and to ensure that $F_+ \cap F_- = \emptyset$ we insist that $\alpha \neq \pm 1$. Note that $[F_+, F_-] \subset \sigma^1$, hence F_+ is a horizontal f-structure on the homogeneous space:

$$\mathbf{g} = \sigma^1 \oplus \bigoplus_{\alpha \in \Delta \setminus \{1\}} \sigma^{\alpha}.$$

§3 : Example: SU(n) flag manifolds

The ideas of this chapter may be illustrated by considering the flag manifolds of SU(n). Invariant f-structures on SU(n) flags of height r are in bijective correspondence with digraphs (directed graphs) on r+1 vertices (see [BS] for more details). In a similar way each invariant metric corresponds to associating a weight (i.e. a strictly positive scalar) to each distinct pair of vertices.

In this context Lemma 4.2 becomes:

Lemma 4.5. A metric Λ and f-structure F on an SU(n) flag manifold satisfies $d^{\nabla}F^{1,1} = 0$ iff:

(i) whenever the configurations

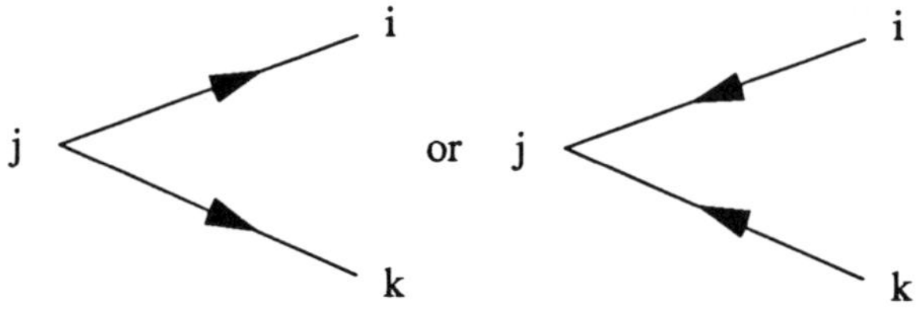

occur, then $\lambda_{ij} = \lambda_{jk}$

and

(ii) whenever the configuration

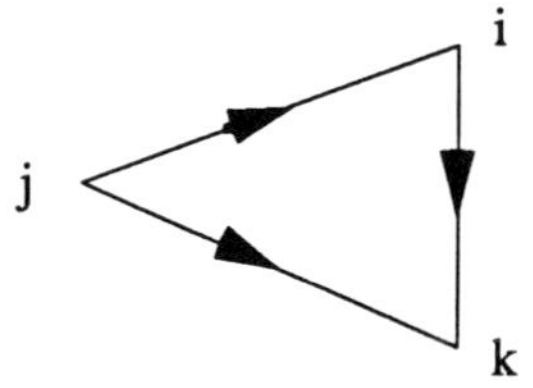

occurs then $\lambda_{jk} = \lambda_{ji} + \lambda_{ik}$.

Note 1. This lemma shows that the condition $(d^{\nabla}F)^{1,1} = 0$ is strictly weaker than 'condition A' of Rawnsley [R2]. 'Condition A' prohibits any configurations of the form

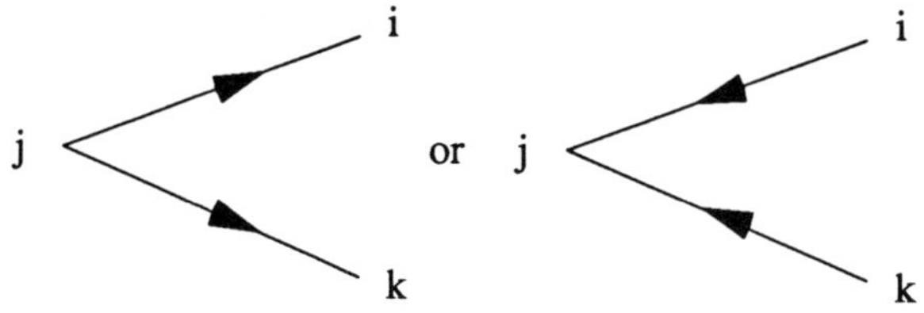

Note 2 The stronger condition of horizontality is that each vertex has at most one edge directed towards it, and at most one edge directed away from it.

Chapter 5
Equi-Harmonic Maps

Prompted in part by the Corollary 4.4, we investigate the general properties of equi-harmonic maps.

Definition. A map into a homogeneous space $^{G}/_{H}$ with a non-empty set of G invariant metrics will be said to be *equi-harmonic* if it is harmonic with respect to each of the G invariant metrics on $^{G}/_{H}$.

Lemma 5.1. Let σ be an automorphism of G which fixes the closed subgroup H. Then σ induces a diffeomorphism $\hat{\sigma}$ of $^{G}/_{H}$. Let φ be a map into $^{G}/_{H}$, then φ is equi-harmonic iff $\hat{\sigma} \circ \varphi$ is equi-harmonic.

Proof. Let Λ be an invariant metric on $^{G}/_{H}$, then $\hat{\sigma} : (^{G}/_{H}, \hat{\sigma} * \Lambda) \to (^{G}/_{H}, \Lambda)$ is an isometric diffeomorphism. Thus φ is harmonic with respect to $\hat{\sigma} * \Lambda$ iff $\hat{\sigma} \circ \varphi$ is harmonic with respect to Λ. The result now follows from the fact that pull back by $\hat{\sigma}$ induces a bijection on the set of invariant metrics. □

Now suppose that G has a bi-invariant metric so that Nöther's Theorem (2.12) applies. We prove an equi-harmonic version of Nöther's Theorem.

Theorem 5.2. Suppose $^G/_H$ is completely reductive, choose a reductive splitting:

$$\mathbf{g} = \mathbf{h} \oplus \mathbf{m}, \quad \mathbf{m} = \bigoplus_{\alpha \in A} \mathbf{m}_\alpha ;$$

and let P_α denote projection onto $\mathbf{m}_\alpha$. Let φ be a mapping of (N,h) into $^G/_H$.
If φ is equi-harmonic, then

$$d^* \varphi^* P_\alpha \circ \beta = 0 \qquad \forall \alpha \in A. \tag{5.1}$$

Conversely, if the reductive splitting has distinct irreducible summands and (1) holds, then φ is equi-harmonic.

Proof. Section 2.4 shows that

$$\sum_{\alpha \in A} \lambda_\alpha P_\alpha \circ \beta \tag{5.2}$$

is the momentum map of an invariant metric for all choices of $\lambda_\alpha > 0$.
Suppose that φ is equi-harmonic.
Nöther's Theorem (2.12) implies that

$$\sum_{\alpha \in A} \lambda_\alpha d^* \varphi^* P_\alpha \beta = 0. \tag{5.3}$$

Taking linear combinations of the versions of equation (5.3) generated by different choice of invariant metric shows that (5.1) holds.

Under the extra hypothesis that $^G/_H$ has distinct irreducible summands, we know that equation (5.2) describes the momentum maps of all the invariant metrics. Hence to establish equi-harmonicity it is sufficient to show that (5.3) holds for all choices of λ_α, this follows trivially from (5.1). □

The next result shows that homogeneous projections allow equi-harmonic maps to proliferate.

Lemma 5.3. Let φ be an equi-harmonic map into $G/_H$, with H strongly reductive. Let K be a closed Lie subgroup of G containing H such that $G/_K$ is completely reductive. The induced map $\pi : G/_H \to G/_K$ is called a *homogeneous projection.* Then $\pi \circ \varphi$ is equi-harmonic.

Proof. Let Λ be an invariant metric on $G/_K$. We can choose a reductive splitting

$$\mathbf{g} = \mathbf{k} \oplus \mathbf{l}, \quad \mathbf{l} = \bigoplus_{\alpha \in A} \mathbf{l}_\alpha ;$$

such that Λ is defined by equation 2.4. The subspaces $\mathbf{l}_\alpha$ are K invariant, hence H invariant. H is strongly reductive so we can choose an H irreducible splitting of $\mathbf{l}_\alpha$ for each $\alpha \in A$:

$$\mathbf{l} = \bigoplus_{\alpha \in A} \bigoplus_{\gamma \in B(\alpha)} \mathbf{l}_{\alpha\gamma}, \qquad \mathbf{l}_\alpha = \bigoplus_{\gamma \in B(\alpha)} \mathbf{l}_{\alpha\gamma}.$$

Let β_1 denote the momentum map on $G/_H$ and let β_2 denote the momentum map on $G/_K$. Theorem 5.2 implies that since φ is equi-harmonic

$$d^* \varphi^* P_{\alpha\gamma} \circ \beta_1 = 0 \qquad \forall \gamma \in B(\alpha), \alpha \in A.$$

However,

$$\pi^* (P_\alpha \circ \beta_2) = \sum_{\gamma \in B(\alpha)} P_{\alpha\gamma} \circ \beta_1$$

and hence

$$d^* (\pi \circ \varphi)^* P_\alpha \circ \beta_2 = \sum_{\gamma \in B(\alpha)} d^* \varphi^* P_{\alpha\gamma} \circ \beta_1 \tag{5.4}$$

$$= 0.$$

Since the momentum map for Λ is of the form 2.4, Nöther's Theorem 2.12 and (5.4) prove that $\pi \circ \varphi$ is harmonic with respect to Λ. □

Lemma 5.3 coupled with Corollary 4.4 provides many twistor results for homogeneous spaces.

Lemma 5.4. Let φ be a map into $G/_H$, with H strongly reductive and such that the tangent space of $G/_H$ has distinct irreducible summands, which is f-holomorphic with respect to a horizontal f-structure. Let K be a closed Lie subgroup of G containing H such that $G/_K$ is completely reductive. Let π be the homogeneous projection $\pi : G/_H \to G/_K$. Then $\pi \circ \varphi$ is equi-harmonic.

Note. The rather involved hypotheses on H and K are actually fairly easy to satisfy. For example, it is sufficient to take G to be compact and choose any H containing a maximal torus of G.

Chapter 6
Classification of Horizontal f-Structures on Flag Manifolds

In this chapter we classify explicitly the invariant horizontal f-structures on flag manifolds. In the final section we state the classification theorem and present some examples of such f-structures which have previously occurred in the literature. Throughout this chapter we will use the Lie group strucure theory set out in §2.6.

§1. Algebraic Preliminaries

Let G be a compact, connected, semi-simple Lie group with sub-torus S and C(S) the centralizer of S in G. A G-flag manifold is the homogeneous space $G/_{C(S)}$. Let $g \in C(S)$, then $\{g\} \cup S$ lies in a maximal torus of G, which is connected; hence C(S) is connected. It follows that flag manifolds are determined by their infinitesimal structure. The isotropy group C(S) is compact and therefore $G/_{C(S)}$ is a completely reductive homogeneous space, in the sense of §2.2.

We denote the Lie algebra of a group by the corresponding lower case boldface.

Flag manifolds have a convenient description in terms of roots, a proof of which may be found in Humphreys [Hu]. The following theorem is essentially the version of that result which appears in [BR2].

Theorem 6.1. Let $\mathbf{t}$ be a maximal torus for $\mathbf{g}$, and $\alpha_1,\dots,\alpha_\ell$ a set of simple roots with respect to a positive root system in $\Delta(\mathbf{g}^{\mathbb{C}}, \mathbf{t}^{\mathbb{C}})$. Then each subset I of $\{1,\dots,\ell\}$ defines a height function n_I on Δ by

$$n_I(\alpha) = \sum_{i \in I} n_i ,$$

where

$$\alpha = \sum_{i=1}^{\ell} n_i \alpha_i .$$

Also, setting $J = \{1,\dots,\ell\}\backslash I$,

$$A(J) = \mathbf{t}^{\mathbb{C}} \oplus \sum_{n_I(\alpha)=0} \mathbf{g}^{\alpha} \qquad \text{(definition)}$$

$$= c(\mathbf{s})^{\mathbb{C}};$$

where we can choose

$$\mathbf{s} = z(\mathbf{h}) = \{\alpha_i, i \in J\}^{\perp}:$$

the centre of $\mathbf{h}$ and the annihilator of $\{\alpha_i, i \in J\}$ in $\mathbf{t}$ respectively.

Moreover, for every sub-torus $\mathbf{u}$ of $\mathbf{g}$, there is an inner automorphism of $\mathbf{g}$ and a subset J of $\{1,\dots,\ell\}$ such that $c(\mathbf{u})^{\mathbb{C}}$ is conjugate to $A(J)$. □

If $c(\mathbf{s})^{\mathbb{C}} = A(J)$ for some $J \subset \{1,\dots,\ell\}$, we will say that $c(\mathbf{s})$ is in *standard form* with respect to the simple roots α_i.

A theorem of Kostant which appears in [Wolf] now allows us to describe the infinitesimal structure of $G/C(S)$ as a $C(S)$ space. Set $H = C(S)$.

Define an equivalence relation $\sim$ on $\Delta(\mathbf{g}^{\mathbb{C}}, \mathbf{t}^{\mathbb{C}})$ by:

$$\alpha \sim \beta \Leftrightarrow \alpha - \beta = \Sigma n_i \gamma_i,\ n_i \in \mathbb{Z},\ \gamma_i \in \Delta(\mathbf{h}^{\mathbb{C}}, \mathbf{t}^{\mathbb{C}}),$$

and let $[\alpha]$ denote the equivalence class containing α. The $\mathbf{h}^{\mathbb{C}}$ module

$$\mathbf{m}_{[\alpha]} = \sum_{\beta \in [\alpha]} \mathbf{g}^{\beta},$$

is $\mathbf{h}^{\mathbb{C}}$ irreducible for $\alpha \notin \Delta(\mathbf{h}^{\mathbb{C}}, \mathbf{t}^{\mathbb{C}})$ and $\mathbf{m}_{[\alpha]}$ is inequivalent to $\mathbf{m}_{[\beta]}$ whenever $[\alpha] \neq [\beta]$.

In the language of Chapter 2 we have a unique reductive splitting

$$T_e^{\mathbb{C}} G/H \cong \mathbf{m}^{\mathbb{C}} \cong \left\{\Sigma\, \mathbf{m}_{[\alpha]} : [\alpha] \in \frac{\Delta(\mathbf{g}^{\mathbb{C}}, \mathbf{t}^{\mathbb{C}})}{\sim}\right\}.$$

Note. Chapter 2 shows that since distinct $\mathbf{m}_{[\alpha]}$ are mutually inequivalent, any $\mathbf{h}^{\mathbb{C}}$ invariant subspace of $\mathbf{m}^{\mathbb{C}}$ is a direct sum of the $\mathbf{m}_{[\alpha]}$.

§2. Reduction to the Irreducible Case

This section concerns itself with the decomposition of an f-structure into 'irreducible' pieces, while Sections 3 and 4 deal with the explicit characterization of such pieces.

First we decompose F_+, as finely as possible, into mutually commuting components. Lemmas 6.2 and 6.3 show that the sub-algebras generated by these components can be dealt with in isolation in that they are mutually orthogonal, commuting sub-algebras of $\mathbf{g}$ and furthermore are compatible with the flag manifold structure.

In Section 1 we saw that flag manifolds are strongly reductive homogeneous spaces with distinct reductive summands. The discussion in Chapter 4 then proves that there is a bijective correspondence between G invariant, horizontal f-structures on G/H and subsets F_+ of $\mathbf{m}^{\mathbb{C}}$ which satisfy:

$$F_+ \cap \overline{F_+} = \{0\}, \tag{f1}$$

$$F_+ \text{ is } \mathbf{h}^{\mathbb{C}} \text{ invariant}, \tag{f2}$$

$$[F_+, \overline{F_+}] \subset \mathbf{h}^{\mathbb{C}}. \tag{f3}$$

Here condition (f3) is that of 'horizontality' of Chapter 4.

It will be notationally convenient to designate F_+ by F^1 in this chapter. We also fix a choice $\mathbf{t}$ of maximal torus for $\mathbf{g}$ contained in $\mathbf{h}$.

F^1 is $\mathbf{h}^{\mathbb{C}}$ invariant, therefore there exists a subset A of $\dfrac{\Delta(\mathbf{g}^{\mathbb{C}}, \mathbf{t}^{\mathbb{C}})}{\sim}$ such that:

$$F^1 = \sum_{[\alpha] \in A} \mathbf{m}_{[\alpha]}.$$

We choose the finest partition of A with the property that if $\alpha, \beta \in \Delta(\mathbf{g}^{\mathbb{C}}, \mathbf{t}^{\mathbb{C}})$ such that $[\alpha], [\beta] \in A$ and $\langle\alpha, \beta\rangle \neq 0$, then $[\alpha]$ and $[\beta]$ lie in the same part of A. Write this partition as

$$A = A_1 \cup \ldots \cup A_r; \qquad A_i \text{ mutually disjoint.}$$

The partition of A induces a decomposition of F^1:

$$F^1 = \sum_{i=1}^{r} F_i^1; \quad F_i^1 = \{\Sigma\, \mathbf{m}_{[\alpha]} : [\alpha] \in A_i\}$$

Definition 6.1.

$$F_i^0 = [F_i^1, \overline{F_i^1}]$$

$$F_i^r = [F_i^1, F_i^{r-1}], \quad r \geq 2$$

$$F_i^{-r} = \overline{F_i^r}, \quad r \geq 1.$$

Lemma 6.2.

(i) F_i^r is an $\mathbf{h}^{\mathbb{C}}$ invariant subspace of $\mathbf{g}^{\mathbb{C}}$.

(ii) $[F_i^r, F_j^s] = 0, \quad i \neq j$.

(iii) $[F_i^r, F_i^s] \subset F_i^{r+s}$.

(iv) F_i^r is orthogonal to F_j^s with respect to the Killing form, B, whenever $i \neq j$.

Proof

(i) F_i^1 is $\mathbf{h}^{\mathbb{C}}$ invariant, now use Definition 6.1.

(ii) Let $\alpha, \beta \in \Delta(\mathbf{g}^{\mathbb{C}}, \mathbf{t}^{\mathbb{C}})$ such that $[\alpha] \in A_i$, $[\beta] \in A_j$ and $i \neq j$. If $\alpha - \beta$ is a root, then by (f3) $\alpha - \beta \in \Delta(\mathbf{h}^{\mathbb{C}}, \mathbf{t}^{\mathbb{C}})$ hence $\beta \in A_i$ which is a contradiction. On the other hand, by the definition of the partition of A, $\langle\alpha, \beta\rangle = 0$ and hence $\alpha + \beta$ is not a root.
So we have proved that if $i \neq j$ then $[F_i^1, F_j^{-1}] = 0$ and $[F_i^1, F_j^1] = 0$.

The statement (ii) now follows from the definition of F_i^r and an inductive application of the Jacobi identity.

(iii) (f3) ensures that $F_i^0 \subset \mathbf{h}^{\mathbb{C}}$. F_i^s is $\mathbf{h}^{\mathbb{C}}$ invariant and together with (i) this proves the case $r = 0$. Induction using the Jacobi identity deals with the remaining cases.

(iv) From Definition 6.1, F_i^1 is orthogonal to F_j^1 and F_j^{-1}. Let $X \in F_i^r$; $Y, Z \in F_j^1$. Then

$$\begin{aligned} B(X, [Y, \bar{Z}]) &= B([X, Y], \bar{Z}) \\ &= 0 \qquad\qquad \text{by (ii).} \end{aligned}$$

Hence $F_i^r \perp F_j^0$.

Let $X \in F_i^r$, $Y \in F_j^1$, $Z \in F_j^{s-1}$, $s \geq 2$. Then

$$B(X, [Y, Z]) = B([X, Y], Z)$$

$$= 0 \qquad \text{by (ii).}$$

The definition of F_j^s shows that $F_i^r \perp F_j^s$. The remaining case $s \leq -2$ follows by conjugation. □

Definition 6.2.

$$\mathbf{a}_i = \sum_{r \in \mathbb{Z}} F_i^r,$$

$$\mathbf{h}_i = \mathrm{Re}\,(\mathbf{a}_i).$$

Lemma 6.3.

(i) $\mathbf{a}_i$ is a Lie sub-algebra of $\mathbf{g}^{\mathbb{C}}$ and $\mathbf{h}_i$ is a Lie sub-algebra of $\mathbf{g}$, both of which are $\mathbf{h}$ invariant.

(ii) $\mathbf{a}_i = \mathbf{h}_i^{\mathbb{C}}$

(iii) $[\mathbf{a}_i, \mathbf{a}_j] = 0$ and $\mathbf{a}_i \perp \mathbf{a}_j$, $i \neq j$.

(iv) $\mathbf{h} \cap \mathbf{h}_i^{\perp}$ acts trivially on $\mathbf{h}_i$.

(v) $\mathbf{h}_i$ is simple.

(vi) $\mathbf{h}^{\mathbb{C}} \cap \mathbf{h}_i^{\mathbb{C}} = F_i^0$.

(vii) $\mathbf{t}_i = \mathbf{t} \cap \mathbf{h}_i$ is a maximal torus for $\mathbf{h}_i$.

(viii) Let $\mathbf{s} = z(\mathbf{h})$ then $\mathbf{h} \cap \mathbf{h}_i = c(\mathbf{s} \cap \mathbf{h}_i) \cap \mathbf{h}_i$.

Proof.

(i), (ii), (iii) follow by applying the appropriate parts of Lemma 6.2.

(iv) Let $X \in \mathbf{h} \cap \mathbf{h}_i^{\perp}$; $Y, Z \in \mathbf{h}_i$. Then

$$B([X, Y], Z) = B(X, [Y, Z])$$
$$= 0,$$

and because $X \in \mathbf{h}_i^{\perp}$ and $\mathbf{h}_i$ is a sub-algebra. $\mathbf{h}_i$ is $\mathbf{h}$ invariant, therefore $[X, Y] \in \mathbf{h}_i$. The Killing form B is non-degenerate, thus $[X, Y] = 0$.

(v) Note that

$$\mathbf{h} = \mathbf{h} \cap \mathbf{h}_i \oplus \mathbf{h} \cap \mathbf{h}_i^{\perp} . \qquad (\dagger)$$

This equation follows because both $\mathbf{h}$ and $\mathbf{h}_i$ are $\mathbf{t}$ invariant, $\mathbf{h}$ contains $\mathbf{t}$, and by equation (6.1) below.

Suppose $\mathbf{h}_i$ is not simple, then $\mathbf{h}_i$ splits properly and $\mathbf{h}_i$ invariantly as:

$$\mathbf{h}_i = \mathbf{h}_{i1} \oplus \mathbf{h}_{i2}; \quad [\mathbf{h}_{i1}, \mathbf{h}_{i2}] = 0, \quad [\mathbf{h}_{ij}, \mathbf{h}_{ij}] \subset \mathbf{h}_{ij} \quad j=1,2 . \qquad (\ddagger)$$

This splitting is automatically $\mathbf{h}$ invariant by (†) and part iv). Furthermore, $\mathbf{h}_i^{\mathbb{C}}$ is generated by F_i^1 so that we obtain a proper splitting

$$F_i^1 = F_{i1}^1 \oplus F_{i2}^1; \; F_{i1}^1 \subset \mathbf{h}_{i1}, \; F_{i2}^1 \subset \mathbf{h}_{i2} .$$

F_{ij}^1 is $\mathbf{h} \cap \mathbf{h}_i$ invariant, hence by (iv) and (†) it is $\mathbf{h}$ invariant. Let $\alpha_j \in \Delta(\mathbf{g}^{\mathbb{C}}, \mathbf{t}^{\mathbb{C}})$ such that $\mathbf{g}^{\alpha_j} \subset F_{ij}^1$, then (‡) proves that $\langle\alpha_1, \alpha_2\rangle = 0$. Referring back to the splitting of F^1 defined on page 42 we see we have a contradiction to the fact that our original partition was as fine as possible, whence (v) follows.

(vi) $F_i^0 \subset h^{\mathbb{C}} \cap h_i^{\mathbb{C}}$, by (f3) and Definition 6.2. $h^{\mathbb{C}} \cap h_i^{\mathbb{C}}$ is a subalgebra of $h_i^{\mathbb{C}}$ containing $t_i^{\mathbb{C}}$. $t_i^{\mathbb{C}}$ is a maximal torus for $h_i^{\mathbb{C}}$ and hence for $h^{\mathbb{C}} \cap h_i^{\mathbb{C}}$. Choose $\alpha \in \Delta(h^{\mathbb{C}} \cap h_i^{\mathbb{C}}, t_i^{\mathbb{C}})$. F_i^1 and F_i^{-1} generate $h_i^{\mathbb{C}}$ and so if X_α commutes with both these sets, X_α would lie in the centre of h_i and hence contradict (v). So we see there exists $\beta \in \Delta(h_i^{\mathbb{C}}, t_i^{\mathbb{C}})$ such that $g\beta \subset F_i^{\pm 1}$ and $\alpha + \beta \in \Delta(h_i^{\mathbb{C}}, t_i^{\mathbb{C}})$.

Now, $F_i^{\pm 1}$ are $h^{\mathbb{C}}$ invariant thus $g^{\alpha+\beta} \subset F_i^{\pm 1}$ and hence Lemma 6.2 (iii) proves that $g^\alpha \subset F_i^0$ and that $[g^\alpha, g^{-\alpha}] \subset F_i^0$. This is sufficient to prove (vi).

(vii) h_i is h invariant, hence t invariant ($t \subset h$). Structure theory shows that any t invariant subspace of g splits as an orthogonal direct sum of a subspace of t with a set of root spaces. Hence

$$h_i = h_i \cap t \oplus h_i \cap t^\perp,$$

from which it follows that

$$t = h_i \cap t \oplus h_i^\perp \cap t. \tag{6.1}$$

$t \cap h_i$ is a torus in h_i. Let t' be a maximal torus for h_i containing $t \cap h_i$. Then $t \cap h_i$ acts trivially on t', by the adjoint action, but $t \cap h_i^\perp$ also acts trivially on t' because it is a subset of $h \cap h_i^\perp$. Applying (6.1) we see that $t' + t$ is a torus for g, however t is maximal and hence $t' = t \cap h_i$.

(viii) Since t_i is a maximal torus for h_i and $t_i \subset t$, we may regard $\Delta(h_i^{\mathbb{C}}, t_i^{\mathbb{C}})$ as a subset of $\Delta(g^{\mathbb{C}}, t^{\mathbb{C}})$ by extending the roots in $(t_i^{\mathbb{C}})^*$ by zero to act on $t^{\mathbb{C}}$.

Apply Theorem 6.1 to see that there is a choice of basis for $\Delta(\mathbf{g}^{\mathbb{C}}, \mathbf{t}^{\mathbb{C}})$ such that $\mathbf{h}$ is in standard form. In the notation of Theorem 6.1, let

$$J_1 = J \cap \Delta(\mathbf{h}_i^{\mathbb{C}}, \mathbf{t}_i^{\mathbb{C}}),$$

$$J_2 = J \backslash J_1 .$$

<u>Claim</u> Let $\alpha \in \Delta(\mathbf{h}_i^{\mathbb{C}}, \mathbf{t}_i^{\mathbb{C}})$ and $j \in J_2$, then $\langle\alpha, \alpha_j\rangle = 0$. Equivalently,

$$\mathbf{t}_i \subset \{\alpha_j, j \in J_2\}^{\perp}.$$

<u>Proof</u> Suppose not, then (without loss of generality) $\alpha + \alpha_j$ is a root. By definition α_j centralizes $\mathbf{s}$ and $\mathbf{h}_i^{\mathbb{C}}$ is $\mathbf{h}^{\mathbb{C}} = c(s)^{\mathbb{C}}$ invariant. Thus $\alpha + \alpha_j \in \Delta(\mathbf{h}_i^{\mathbb{C}}, \mathbf{t}_i^{\mathbb{C}})$, but $\mathbf{h}_i^{\mathbb{C}}$ is a subalgebra so:

$$(\alpha + \alpha_j) - \alpha = \alpha_j \in \Delta(\mathbf{h}_i^{\mathbb{C}}, \mathbf{t}_i^{\mathbb{C}}),$$

which is a contradiction to the definition of J_2.

The second statement follows because $\mathbf{h}_i$ is simple and thus $\mathbf{t}_i$ is generated by linear combinations of $\{ H_\alpha : \alpha \in \Delta(\mathbf{h}_i^{\mathbb{C}}, \mathbf{t}_i^{\mathbb{C}}) \}$.

An immediate consequence of the claim is that

$$\mathbf{s} \cap \mathbf{h}_i = \mathbf{s} \cap \mathbf{t}_i = \mathbf{t}_i \cap \{\alpha_j, j \in J_1\}^{\perp}.$$

Thus,

$$\mathbf{t}_i = \mathbf{s} \cap \mathbf{t}_i \oplus \langle H_{\alpha_j} : j \in J_1\rangle, \qquad (*)$$

where $\langle\, , \rangle$ denotes the linear subspace generated by the elements enclosed.

Clearly $c(\mathbf{s}) \cap \mathbf{h}_i \subset c(\mathbf{s} \cap \mathbf{h}_i) \cap \mathbf{h}_i$ and both these subalgebras intersect $\mathbf{t}$ in $\mathbf{t}_i$. Orthogonal to $\mathbf{t}$ they consist of a direct sum of root spaces in $\Delta(\mathbf{g}^{\mathbb{C}}, \mathbf{t}^{\mathbb{C}})$. Choose $\mathbf{g}^{\alpha} \subset c(\mathbf{s} \cap \mathbf{h}_i) \cap \mathbf{h}_i$, then $\alpha \in \Delta(\mathbf{h}_i^{\mathbb{C}}, \mathbf{t}_i^{\mathbb{C}})$ so $H_\alpha \in \mathbf{t}_i$ and further $H_\alpha \perp \mathbf{s} \cap \mathbf{t}_i$. Hence (*) shows that $H_\alpha \in \langle H_{\alpha_j} : j \in J_1 \rangle$, so the definition of $\mathbf{s}$ proves $H_\alpha \perp \mathbf{s}$. Said otherwise, $\mathbf{g}^{\alpha} \subset c(\mathbf{s})$. □

Lemma 6.4 $$F_i^1 \cap F_i^s \neq \{0\} \Rightarrow F_i^1 \subset F_i^s.$$

Proof Let

$$A_i^s = \{ [\alpha] : \mathbf{g}^{\gamma} \subset F_i^1 \cap F_i^s \quad \forall\, \gamma \in [\alpha] \}.$$

Clearly $A_i^s \subset A_i$, and since $F_i^1 \cap F_i^s$ is $\mathbf{h}$ invariant

$$F_i^1 \cap F_i^s = \{\Sigma\, \mathbf{g}^{\alpha} \mid [\alpha] \in A_i^s \}.$$

Furthermore, suppose there exists $\beta \in \Delta(\mathbf{g}^{\mathbb{C}}, \mathbf{t}^{\mathbb{C}})$ such that $[\beta] \in A$ and $\langle\alpha,\beta\rangle \neq 0$ for some α with $[\alpha] \in A_i^s$.

Then

$$\begin{aligned}[X_\beta, [X_{-\alpha}, X_\alpha]] &= \beta([X_{-\alpha}, X_\alpha])X_\beta \\ &= \langle\alpha, \beta\rangle X_\beta .\end{aligned}$$

Notice that $[X_{-\alpha}, X_\alpha] \in F_i^{s-1}$ (Lemma 6.2 (iii)), hence $\langle\alpha, \beta\rangle X_\beta \in F_i^s$ by the same result, $\langle\alpha, \beta\rangle$ is non-zero, so we see that $g^{\beta} \subset F_i^s$, but F_i^s is $\mathbf{h}$ invariant and so all the root spaces in $[\beta]$ lie in F_i^s.

So A_i^s satisfies the properties required of the parts of the partition of page 42 however this was already chosen to be as fine as possible and therefore $A_i^s = A_i$. Equivalently, $F_i^s \cap F_i^1 = F_i^1$. □

To summarize, we have obtained a description of a horizontal f-structure on a G-flag manifold as a direct sum of horizontal f-structures on flag manifolds of mutually orthogonal, commuting subgroups H_i of G each satisfying the following additional hypotheses.

$\mathbf{h}$ is simple. (s1)

$c(\mathbf{s})^{\mathbb{C}} = [F^1, F^{-1}]$. (s2)

F^1 and F^{-1} generate $\mathbf{h}^{\mathbb{C}}$. (s3)

We say that an f-structure is *irreducible* if, in the notation above, p = 1 and $\mathbf{h}_1 = \mathbf{g}$. This section shows that in order to obtain a general classification of horizontal f-structures, it is sufficient to consider the irreducible case. In sections 3 and 4 we therefore assume that F is irreducible: ie. F satisfies (f1) - (f3) and (s1) - (s3).

Property (s3) subdivides:

F^1 generates $\mathbf{g}^{\mathbb{C}}$, (s3′)

or

F^1 and F^{-1} generate $\mathbf{g}^{\mathbb{C}}$, but F^1 alone does not. (s3″)

Sections 4 discusses the former option, section 3 the latter.

§3. Characterization of Irreducible f-structures I

In this section we assume that F is irreducible and satisfies

$$(f1), (f2), (f3), (s1), (s2), \text{ and } (s3'').$$

Define a subset Φ^+ of $\Delta(g^{\mathbb{C}}, t^{\mathbb{C}})$ by:

$$\alpha \in \Phi^+ \Leftrightarrow g^\alpha \subset F^s, \; s > 0.$$

Then, from Lemma 6.2(iii)

$$\Phi^+ + \Phi^+ \subset \Phi^+. \tag{6.2}$$

Claim

$$\Phi^+ \cap - \Phi^+ = \emptyset. \tag{6.3}$$

Proof. Suppose not. Then there is an identity

$$\alpha_{i_1} + \ldots + \alpha_{i_r} = - \alpha_{j_1} - \ldots - \alpha_{j_s} \qquad r,s \geq 1$$

with $\alpha_k \in \Phi^+$ and each partial sum

$$\alpha_{i_1} + \ldots + \alpha_{i_t}, \qquad 1 \leq t \leq r;$$

$$-\alpha_{j_1} - \ldots - \alpha_{j_t}, \qquad 1 \leq t \leq s$$

is a root.

We may re-arrange this to give:

$$\alpha_{i_1} + \ldots + \alpha_{i_r} + \alpha_{j_s} + \ldots + \alpha_{j_2} = - \alpha_{j_1};$$

and again each partial sum is a root.

In the notation of §6.2

$$\mathbf{g}^{-\alpha_{j_1}} \subset F^{-1} \cap F^{r+s-1},$$

and hence Lemma 6.4 shows that

$$F^{-1} \subset F^{r+s-1}. \tag{6.4}$$

(6.4) implies that F^{-1} lies in the set generated by F^1 which contradicts our assumption (s3″).

Having established the claim (6.3), we may now apply a lemma of Borel and Hirzebruch [BH] to see that Φ^+ may be extended to a positive set of roots Δ^+. This ordering on $\Delta(\mathbf{g}^{\mathbb{C}}, \mathbf{t}^{\mathbb{C}})$ induces an ordering on $\Delta(c(s)^{\mathbb{C}}, \mathbf{t}^{\mathbb{C}})$ with respect to which we have simple roots $\alpha_1,\dots,\alpha_\ell$. Number the $c(s)$–irreducible subspaces of F^1 and let

$\alpha_{\ell+i}$ = lowest root in the ith $c(s)$–irreducible subspace of F^1.

Lemma 6.5. $\{\alpha_1,\dots,\alpha_\ell, \alpha_{\ell+1},\dots,\alpha_k\}$ are a set of simple roots for Δ^+.

Proof. By definition, $\{\alpha_1,\dots,\alpha_\ell\}$ generates the positive root spaces in $c(s)^{\mathbb{C}}$. Since $\alpha_{\ell+i}$ is the lowest weight in the ith $c(s)$ irreducible subspace of F^1, the action of $\{\alpha_1,\dots,\alpha_\ell\}$ on $\{\alpha_{\ell+1},\dots,\alpha_k\}$ generates F^1. F^1 generates the positive root spaces lying outside of $c(s)^{\mathbb{C}}$. Hence any positive root β in $\Delta(\mathbf{g}^{\mathbb{C}}, \mathbf{t}^{\mathbb{C}})$ may be written

$$\beta = \sum_{i=1}^{k} n_i\, \alpha_i, \quad n_i \geq 0. \tag{6.5}$$

Claim. If $i \neq j$ then $\alpha_i - \alpha_j$ is not a root.

Proof of Claim. If $1 \leq i \neq j \leq \ell$, then the result follows because both α_i and α_j are simple roots in $\Delta(c(s)^{\mathbb{C}}, \mathbf{t}^{\mathbb{C}})$.

If $\ell+1 \le i \le k$, $1 \le j \le \ell$ then the result holds because α_i is the lowest weight of a representation of c(s).

If $\ell+1 \le i \ne j \le k$ then by (f3); if $\alpha_i - \alpha_j$ is a root, then it lies in $\Delta(c(s)^{\mathbb{C}}, \mathfrak{t}^{\mathbb{C}})$ in which case α_i and α_j lie in the same c(s) irreducible subspace of $\mathbf{g}^{\mathbb{C}}$ - which is a contradiction. This claim is now proved.

An immediate consequence of the claim is that

$$\langle\alpha_i, \alpha_j\rangle \le 0 \quad i \ne j. \tag{6.6}$$

Suppose α_s is not simple, then

$$\alpha_s = \beta + \gamma \quad \text{with} \quad \beta, \gamma \in \Delta^+.$$

From (6.5)

$$\alpha_s = \Sigma\, n_i\, \alpha_i, \quad n_i \ge 0.$$

If $n_s = 0$, taking the inner product of (6.5) with α_s contradicts (6.6). Whence there is an identity:

$$\Sigma\, m_i\, \alpha_i = 0, \; m_i \ge 0, \; \text{not all } m_i = 0. \tag{6.7}$$

Take the inner product of (6.7) with α_j and apply (6.6) to see that $m_j = 0$. Again we have a contradiction, and so α_s is a simple root. □

We summarize the results of this section in a theorem:

Theorem 6.6. Let F satisfy (f1) – (f3), (s1), (s2) and (s3″). Then we may choose a set of simple roots Δ for $\mathbf{g}$ and a subset Δ' of Δ such that F is in standard form: i.e.

(i) $c(s)^{\mathbb{C}} = A(\Delta')$

(ii) $F^1 = \{\Sigma\, \mathbf{g}^{\alpha} \mid \alpha \in \Delta(\mathbf{g}^{\mathbb{C}}, \mathfrak{t}^{\mathbb{C}}),\; n_I(\alpha) = 1\}$ with $I = \Delta \setminus \Delta'$.

§4. Characterisation of Irreducible f-structures II.

In this section we assume that F is irreducible and satisfies

(∗) (f1), (f2), (f3), (s1), (s2), and (s3′).

Note. Because $\mathbf{g}^{\mathbb{C}}$ has trivial centre and F^1 generates $\mathbf{g}^{\mathbb{C}}$,

$$[X, F^1] = 0 \Rightarrow X = 0 \tag{6.8'}$$

and conjugately

$$[X, F^{-1}] = 0 \Rightarrow X = 0. \tag{6.8''}$$

Lemma 6.7. Assume F satisfies (∗), then

(i) $F^s \cap F^r \neq \{0\} \Rightarrow F^s = F^r$.

(ii) $[F^1, F^r] = F^{r+1}$ and $[F^{-1}, F^r] = F^{r-1}$.

Proof

(i) Suppose $F^s \cap F^r \neq \{0\}$, then invoking equations (6.8)

$$\{0\} \neq \underbrace{[F^{-1}, [F^{-1}, \ldots [F^{-1},}_{s-1 \text{ copies}} F^s \cap F^r] \ldots]$$

$$\subset F^1 \cap F^{r-s+1} \qquad \text{by Lemma 6.2(iii).}$$

Lemma 6.4 implies

$$F^1 \subset F^{r-s+1}. \tag{6.9}$$

Case $s \geq 2$. The definition of F^s shows that

$$F^s = [\,\underbrace{F^1,\dots,[\,F^1, F^1}_{\text{s copies}}\,]\dots]$$

$$\subset [\,\underbrace{F^1,\dots,[\,F^1}_{\text{s}-1\text{ copies}}, F^{r-s+1}]\dots] \qquad \text{by (6.9).}$$

$$\subset F^r \qquad \text{by Lemma 6.2(iii).}$$

Case $s = 0$.

$$F^0 = [\,F^{-1}, F^1\,]$$

$$\subset [\,F^{-1}, F^{r+1}\,] \qquad \text{by (6.9),}$$

$$\subset F^r.$$

Case $s = 1$. This is Lemma 6.4.

The case $s < 0$ follows by conjugating the cases $s > 0$. Swapping the roles of r and s completes the proof that $F^r = F^s$.

(ii) When $r \geq -1$, the first statement follows by Definition 6.1. So suppose $r < -1$. F^1 generates $\mathbf{g}^{\mathbb{C}}$, therefore $\exists t > 0$ such that

$$F^t \cap F^r \neq \{0\}.$$

Part (i) implies that $F^t = F^r$. Thus

$$F^{1+t} = [\,F^1, F^t\,] = [\,F^1, F^r\,] \subset F^{1+r}, \tag{6.10}$$

where the last inclusion follows from Lemma 6.2 iii).

Equations (6.8) prove that $F^{1+t} \neq \{0\}$ and part (i) implies that equality holds throughout (6.10) to give the first part of (ii). The second part follows by conjugation. □

Lemma 6.8. If F satisfies (∗) then $\exists m > 2$ such that

$$F^t = F^u \Leftrightarrow t \equiv u \bmod m. \qquad (6.11)$$

Proof. Let m be the smallest strictly positive integer such that

$$F^m \cap F^0 \neq \{0\}.$$

Such m exists because F^1 generates $\mathbf{g}^{\mathbb{C}}$.

If $m = 1$ then $F^1 \cap F^0 \neq \{0\}$ which contradicts the fact that $F^1 \subset \mathbf{m}^{\mathbb{C}}$.

If $m = 2$ then by Lemma 6.7, $F^0 = F^2$ hence $F^1 = F^{-1}$ which contradicts (f1).

⇐] Apply Lemma 6.7 and Definition 6.1 to see that

$$F^{m+r} = F^r \ \forall r \in \mathbb{Z},$$

and hence

$$F^{km} = F^0 \ \forall k \in \mathbb{Z}.$$

Thus

$$F^{km+p} = \underbrace{[F^1, \ldots [F^1}_{p \text{ copies}}, F^{km}] \ldots] = \underbrace{[F^1, \ldots, [F^1}_{p \text{ copies}}, F^0] \ldots].$$

Now, because $F^0 = c(s)^{\mathbb{C}}$, F^1 is a direct sum of F^0 irreducible spaces and therefore $[F^1, F^0] = F^1$. Further,

$$F^{km+p} = \underbrace{[F^1, \ldots, [F^1, F^1}_{p \text{ copies}}] \ldots]$$

$$= F^p \qquad \text{by definition when } p > 0.$$

$\Rightarrow$] Suppose $F^t = F^u$ with $t > u$. Then

$$F^0 = [\, F^t, \overline{F^u}\,] \subset F^{t-u},$$

hence

$$F^{t-u} = F^0.$$

Similar arguments to those in the proof of $\Leftarrow$] show that we have

$$F^{t-u} = F^r \text{ whenever } r = km + (t-u),$$

and we may choose k such that $0 \leq r < m$. The definition of m shows that

$$t - u \equiv 0 \mod m. \qquad \square$$

Lemma 6.9. Assume F satisfies (∗), let m be as defined by Lemma 6.8 and let ε be a primitive m^{th} root of unity. Define $\tau : \mathbf{g}^{\mathbb{C}} \to \mathbf{g}^{\mathbb{C}}$ by:

$$\tau(X) = \varepsilon^k X \text{ for } X \in F^k.$$

Then

(i) τ is well defined and preserves $\mathbf{g}$.

(ii) The ε eigenspace of τ generates $\mathbf{g}^{\mathbb{C}}$.

(iii) τ is an inner automorphism of order m for which the set of fixed points is $c(s)^{\mathbb{C}}$.

Proof.

(i) Lemmas 6.7 and 6.8 show that τ is well defined. $\overline{F^k} = F^{-k}$ proves the second part.

(ii) F satisfies (s3″).

(iii) Definition of F^k and (ii) shows τ is an automorphism. Lemmas 6.7 and 6.8 show that $c(\mathbf{s})^{\mathbb{C}}$ is the fixed point set for τ. A lemma appearing in Helgason[1] states that any automorphism of $\mathbf{g}$ which fixes a maximal torus pointwise is inner. Together with part (i) this gives the result. □

In Section 3, a lemma of Borel and Hirzebruch was sufficient to settle the classification. Here we must digress to set up the use of a more sophisticated tool.

The classification theory of finite order automorphisms of complex simple Lie algebras due to Kac and presented in Helgason[2] allows us to extract the following result, which classifies inner automorphisms.

Theorem 6.10 (Kac). Let $\mathbf{g}$ be a compact, simple Lie algebra over $\mathbb{R}$ with maximal torus $\mathbf{t}$. Choose a set of simple roots $\alpha_1,\dots,\alpha_\ell$ in $\Delta(\mathbf{g}^{\mathbb{C}}, \mathbf{t}^{\mathbb{C}})$ and let α_0 be minus the highest root. Let $(s_0,\dots,s_\ell)$ be integers ≥ 0 without non-trivial common factor, put $m = \sum_0^\ell a_i s_i$, where a_i is the coefficient of α_i in the highest root and $a_0 = 1$. Let ε be a primitive m^{th} root of unity. Then:

(i) The prescription that σ acts by multiplication by ε^{s_i} on $\mathbf{g}^{\alpha_i}$ $(0 \leq i \leq \ell)$ defines uniquely an inner automorphism of $\mathbf{g}^{\mathbb{C}}$ of order m. It will be called an inner automorphism of type $(s_0,\dots,s_\ell)$.

(ii) Except for conjugation, the automorphisms σ exhaust all m^{th} order inner automorphisms of $\mathbf{g}^{\mathbb{C}}$. □

---

1 [He] Chap. IX, 5.3.

2 ibid Chap. X, 5.15, 5.16.

Let σ be an automorphism of $\mathbf{g}^{\mathbb{C}}$ which fixes $\mathbf{t}^{\mathbb{C}}$ pointwise. Then the root spaces of $\mathbf{g}^{\mathbb{C}}$ with respect to $\mathbf{t}^{\mathbb{C}}$ are invariant under σ and thus σ is characterized by its action on the root spaces.

Lemma 6.11 (In the notation of Theorem 6.10). Suppose σ preserves $\mathbf{g}$ and is of type $s = (s_0,\ldots,s_\ell)$. Let $\alpha \in \Delta(\mathbf{g}^{\mathbb{C}}, \mathbf{t}^{\mathbb{C}})$ then α has a unique expression of the form

$$\alpha = \sum_{i=1}^{\ell} n_i\, \alpha_i.$$

Define the s-*height of* α to be

$$h_s(\alpha) = \sum_{i=1}^{\ell} n_i\, s_i.$$

Then σ acts by multiplication by $\varepsilon^{h_s(\alpha)}$ on $\mathbf{g}^{\alpha}$.

Proof. If α is a positive root then

$$\alpha = \alpha_{i_1} + \ldots + \alpha_{i_r}\,, \quad 1 \le i_j \le \ell,$$

with each partial sum a root. The definition of σ in Theorem 6.10 proves the result.

When α is negative, then σ acts by multiplication by $h_s(-\alpha)$ on $\mathbf{g}^{-\alpha}$. Furtheremore, σ preserves $\mathbf{g}$ and therefore acts by multiplication by the conjugate of $\varepsilon^{h_s(-\alpha)}$ on $\mathbf{g}^{\alpha}$, that is by $\varepsilon^{h_s(\alpha)}$. □

We are now able to determine the structure of F.

Theorem 6.12. Let F satisfy (∗) and let τ be as defined in Lemma 6.9.

(i) There is a choice of simple roots Δ of $\Delta(\mathbf{g}^{\mathbb{C}}, \mathbf{t}^{\mathbb{C}})$ such that τ is of type s.

(ii) $c(s)^{\mathbb{C}}$ is in standard form for this choice of Δ, i.e. $c(s)^{\mathbb{C}} = A(\Delta')$ for some $\Delta' \subset \Delta$.

Further,

$$s_i = 0 \Leftrightarrow \alpha_i \in A(\Delta'),$$

$$s_i = 1 \Leftrightarrow \alpha_i \notin A(\Delta'),$$

and hence $h_s(\alpha) = n_I(\alpha)$ where $I = \Delta \setminus \Delta'$.

(iii) $m = n_I(-\alpha_0) + 1$.

(iv) $F^1 = \{ \Sigma\, \mathbf{g}^{\alpha} \mid n_I(\alpha) \equiv 1 \bmod m \}$.

Proof.

(i) follows by elementary arguments from part (ii) of Theorem 6.10.

(ii) τ is of type s, therefore the fixed point set of τ is $\mathbf{t}^{\mathbb{C}}$ together with the root spaces α for which $h_s(\alpha) \equiv 0 \bmod m$.

Now, $m = \sum_{i=0}^{\ell} a_i s_i$, so that

$$-m < h_s(\alpha) < m \quad \forall \alpha \in \Delta(\mathbf{g}^{\mathbb{C}}, \mathbf{t}^{\mathbb{C}}). \tag{6.12}$$

Thus $h_s(\alpha) \equiv 0 \bmod m \quad \Leftrightarrow \quad h_s(\alpha) = 0$

$$\Leftrightarrow \quad \alpha \in A(\Delta') \text{ where } \Delta' = \{\alpha_i \in \Delta \mid s_i = 0\}.$$

Lemma 6.9(iii) shows that

$$s_i = 0 \Leftrightarrow \alpha_i \in A(\Delta').$$

We calculate the ε eigenspace of τ. By Lemma 6.11 this is:

$$\mathbf{g}_\tau^\varepsilon = \{\Sigma\, \mathbf{g}^\alpha \mid h_s(\alpha) \equiv 1 \bmod m\}, \tag{6.13}$$

and equation (6.12) shows that

$$\mathbf{g}_\tau^\varepsilon = B \oplus C;$$

$$B = \{\Sigma\, \mathbf{g}^\alpha \mid h_s(\alpha) = 1\},$$

$$C = \{\Sigma\, \mathbf{g}^\alpha \mid h_s(\alpha) = 1 - m\}.$$

We can describe B immediately:

$$B = \{\Sigma\, \mathbf{g}^\alpha \mid \alpha \sim \alpha_i \bmod A(\Delta'),\ \alpha_i \in \Delta \text{ and } s_i = 1\}.$$

Consider the equation

$$h_s(\alpha) = 1 - m,\quad \alpha = \sum_{i=1}^{\ell} n_i\, \alpha_i.$$

We get

$$a_0\, s_0 + \sum_{i=1}^{\ell} (a_i + n_i)s_i = 1.$$

The a_i are the coefficients of the highest root and thus $a_i + n_i \geq 0$. There are three cases:

(a) $s_0 > 1 \Rightarrow C = \emptyset$,

(b) $s_0 = 1 \Rightarrow C = \{\Sigma\, \mathbf{g}^\alpha \mid \alpha \sim \alpha_0 \bmod A(\Delta')\}$,

(c) $s_0 = 0 \Rightarrow \alpha_0 \in A(\Delta')$ whence $A(\Delta') = \mathbf{g}^{\mathbb{C}}$ since α_0 is the lowest weight and thus $c(s)^{\mathbb{C}} = \mathbf{g}^{\mathbb{C}}$ which is a trivial case.

In the first instance the ε-eigenspace consists only of positive roots. Positive roots are closed under Lie bracket and so cannot generate $\mathbf{g}^{\mathbb{C}}$, this contradicts Lemma 6.9(iii). So we see that we may assume $s_0 = 1$ and retain full generality. Finally, if there is an

$\alpha_i \in \Delta \setminus \Delta'$ for which $s_i > 1$ then $\mathbf{g}_{\tau}^{\varepsilon}$ does not generate $\mathbf{g}^{\mathbb{C}}$ again contradicting Lemma 6.9(iii).

Statements (iii) and (iv) follow easily. □

§5. Summary and Examples

We summarize the results of Sections 2-4 as follows.

Theorem 6.13. Let $\mathbf{g}$ be a compact real Lie algebra with toral sub-algebra $\mathbf{u}$. Let $\mathbf{s} = z(c(\mathbf{u}))$, then $c(\mathbf{s}) = c(\mathbf{u})$. Let F^+ be a subspace of $\mathbf{g}^{\mathbb{C}}$ satisfying:

$$F^+ \cap c(\mathbf{s})^{\mathbb{C}} = \{0\}, \tag{f0}$$

$$F^+ \cap \overline{F^+} = \{0\}, \tag{f1}$$

$$F^+ \text{ is } c(\mathbf{s})^{\mathbb{C}} \text{ invariant}, \tag{f2}$$

$$[F^+, \overline{F^+}] \subset c(\mathbf{s})^{\mathbb{C}}. \tag{f3}$$

Choose a maximal torus $\mathbf{t}$ of $\mathbf{g}$ such that $c(\mathbf{s})$ can be put in standard form with respect to $\mathbf{t}$. Then there exist mutually orthogonal, commuting, simple sub-algebras $\mathbf{h}_i$ $(1 \le i \le p)$ of $\mathbf{g}$ with the following properties.

(i) $F^+ \subset \sum_{i=1}^{p} \mathbf{h}_i^{\mathbb{C}}$.

(ii) $c(\mathbf{s}) \cap \mathbf{h}_i = c(\mathbf{s} \cap \mathbf{h}_i) \cap \mathbf{h}_i$ and $\mathbf{t}_i = \mathbf{t} \cap \mathbf{h}_i$ is a maximal torus for $\mathbf{h}_i$.

(iii) There exists a choice of simple roots Δ_i in $\Delta(\mathbf{h}_i^{\mathbb{C}}, \mathbf{t}_i^{\mathbb{C}})$ and a subset $\Delta_i' \subset \Delta_i$ such that

$$c(\mathbf{s} \cap \mathbf{h}_i)^{\mathbb{C}} = A(\Delta_i')$$

and, setting $I_i = \Delta_i \setminus \Delta_i'$ and $F_i^+ = F^+ \cap \mathbf{h}_i^{\mathbb{C}}$ either:

$$F_i^+ = \{ \Sigma\, \mathbf{g}^{\alpha} \mid \alpha : n_{I_i}(\alpha) = 1 \quad \alpha \in \Delta(\mathbf{h}_i^{\mathbb{C}}, \mathbf{t}_i^{\mathbb{C}}) \} \tag{A}$$

or

$$F_i^+ = \{\Sigma\, \mathbf{g}^\alpha \mid \alpha : n_{I_i}(\alpha) \equiv 1 \bmod n_I(\theta) + 1;\ \ \alpha \in \Delta(\mathbf{h}_i^{\mathbb{C}}, \mathbf{t}_i^{\mathbb{C}}),\ \theta = \text{highest root}\} \qquad \text{(B)}$$

and rank $\mathbf{h}_i > 1$.

Conversely, any subspace F^+ which satisfies (i), (ii) and (iii) also satisfies (f0) - (f3).

Proof. Theorem 6.1 proves the second sentence. Sections 2 to 4 prove everything else except the final sentence and in that case to establish (f0) - (f2) is trivial, while (f3) follows because

$$n_I(\alpha + \beta) = n_I(\alpha) + n_I(\beta) \qquad \text{for } \alpha + \beta, \alpha, \beta \in \Delta(\mathbf{g}^{\mathbb{C}}, \mathbf{t}^{\mathbb{C}}) \cup \{0\}.$$

The theorem shows that F is irreducible iff **g** is simple and F is of the form (A) or (B) of the theorem.

Definition 6.3. An f-holomorphic map into a homogeneous space with invariant f-structure will be said to be *full* if there is no smaller invariant f-structure with respect to which it is f-holomorphic. (Here 'smaller' means having smaller +i eigenspace.)

It is easy to see that, at least for reductive homogeneous spaces with distinct irreducible summands, each f-holomorphic map has a smallest invariant f-structure with respect to which it is f-holomorphic.

Corollary 6.14. Suppose M is connected and $\varphi : M \to {}^{G}/_{C(S)}$ is f-holomorphic with respect to a horizontal f-structure. Then there exists a set H_i $(i = 1,...,p)$ of mutually orthogonal, commuting, simple subgroups of G and an element g of G such that a translate $L_g \circ \varphi$ factors as

$$L_g \circ \varphi : M \to \frac{H_1}{C(S \cap H_1)} \times \ldots \times \frac{H_p}{C(S \cap H_p)} \hookrightarrow \frac{G}{C(S)}$$

and such that the ith component of $L_g \circ \varphi$ is f-holomorphic and full with respect to an irreducible horizontal f-structure on $\dfrac{H_i}{C(S \cap H_i)}$.

Before embarking on the proof of the Corollary we require a Lemma.

Lemma 6.15. Let G be a compact, semi-simple Lie group. Denote the Killing form on $\mathbf{g}$ by $\langle \, , \rangle$. Let H be a connected Lie subgroup and $\mathbf{k}$ a subalgebra of $\mathbf{g}$ such that

$$[\mathbf{h}, \mathbf{k}] \subset \mathbf{k} \qquad (*)$$

and

$$\mathbf{k} = \mathbf{h} \cap \mathbf{k} \oplus \mathbf{h}^{\perp} \cap \mathbf{k} . \qquad (\dagger)$$

Then $\mathbf{h}^{\perp} \cap \mathbf{k}$ is $\mathbf{h}$ invariant and, under the identification of $T\,{}^{G}/_{H}$ with $\mathbf{h}^{\perp}$, defines an involutive G invariant distribution on ${}^{G}/_{H}$.

Set $K = \exp(\mathbf{k})$. The space $\dfrac{K}{K \cap H}$ constitutes an embedded maximal integral submanifold for this distribution, passing through the identity coset of ${}^{G}/_{H}$.

Proof. Notice that because the Killing form is non-degenerate, $\mathbf{h}^{\perp}$ is $\mathbf{h}$ invariant. Hence $\mathbf{h}^{\perp} \cap \mathbf{k}$ is also $\mathbf{h}$ invariant and may thus be extended over G/H to give a G invariant distribution $\underline{\mathbf{k} \cap \mathbf{h}}^{\perp}$.

It will be necessary to distinguish between vector field Lie bracket $\{ , \}$ and the Lie bracket in $\mathbf{g}$, $[,]$.

Lemma 2.3 shows that (suppressing the isomorphism β)

$$T^{c}(X,Y) = -\tfrac{1}{2}P_{\underline{\mathbf{h}}^{\perp}}[X,Y], \qquad (\ddagger)$$

where X and Y are vector fields on G/H. Let $X,Y \in \Gamma(\underline{\mathbf{k} \cap \mathbf{h}}^{\perp})$ and apply $(\ddagger)$ to see that

$$\{X,Y\} = -\tfrac{1}{2}P_{\underline{\mathbf{h}}^{\perp}}[X,Y] - \nabla^{c}_{X}Y + \nabla^{c}_{Y}X .$$

Now $\underline{\mathbf{k} \cap \mathbf{h}}^{\perp}$ is an $\mathbf{h}$ invariant and hence parallel sub-bundle of $\underline{\mathbf{h}}^{\perp}$ for the canonical connection ∇^{c}. Furthermore, the fact that $\mathbf{k}$ is a subalgebra and $(\dagger)$ shows that $P_{\underline{\mathbf{h}}^{\perp}}[X,Y]$ is a section of $\underline{\mathbf{k} \cap \mathbf{h}}^{\perp}$. Thus $\{X,Y\} \in \Gamma(\underline{\mathbf{k} \cap \mathbf{h}}^{\perp})$.

The final statement follows from $(\dagger)$. □

Proof of Corollary 6.14

Let $\mathbf{k} = \sum_{i=1}^{p} \mathbf{h}_i$.

It is clear that $\mathbf{k}$ is a subalgebra of $\mathbf{g}$, and arguments similar to those contained in the proof of Lemma 6.3 show that $\mathbf{k}$ satisfies $(\dagger)$. Since each $\mathbf{h}_i$ is $\mathbf{h}$ invariant, $\mathbf{k}$ is $\mathbf{h}$ invariant. We may thus apply Lemma 6.15 to $\mathbf{k}$ over G/H.

It is now necessary to identify $\dfrac{K}{K \cap H}$. Let $H_i = \exp(\mathbf{h}_i)$, then

$$K = \exp(\mathbf{k}) = \exp\Big(\sum_{i=1}^{p} \mathbf{h}_i\Big) = \prod_{i=1}^{p} H_i \ ;$$

the last element of the identity being the internal direct product in G.

Claim $K \cap H = \prod_{i=1}^{p} (H \cap H_i) = \prod_{i=1}^{p} C_G(S \cap H_i) \cap H_i$

Proof Let $g \in C_G(S \cap H_i) \cap H_i$, then $\{g\} \cup (S \cap H_i)$ lies in a maximal torus of H_i which in turn is contained in $C_G(S \cap H_i)$. It follows that $C_G(S \cap H_i)$ is connected. Lemma 6.3 proves the infinitesimal version of the second identity. On the other hand it is obvious that

$$C(S) \cap H_i \subset C_G(S \cap H_i) \cap H_i,$$

in the light of which we see that these two groups are equal and so the second identity holds.

Consider $H_i \cap H_j$ for $1 \leq i \neq j \leq p$. Since H_i and H_j commute this intersection must lie in the centre of both H_i and H_j; hence in the intersection of their respective maximal tori. Theorem 6.13 states that infinitesimally these tori are orthogonal from which it follows that $H_i \cap H_j$ consists only of the identity element in G.

As a result, the external direct product of the H_i is isomorphic to their internal direct product in G.

Let $g \in K \cap H$, then

$$g = g_1 g_2 \cdots g_p, \quad g_i \in H_i.$$

Let $s \in S$, and recall $H = C(S)$ so:

$$g_1 g_2 \cdots g_p = g = s g s^{-1} = \prod_{i=1}^{p} s g_i s^{-1}.$$

Recalling the remarks of the previous paragraph, it follows that

$$g_i = s g_i s^{-1} \qquad i = 1, \ldots, p,$$

and thus

$$g \in \prod_{i=1}^{p} (H \cap H_i).$$

The claim is now proved. □

The claim establishes that

$$\frac{K}{K \cap H} = \frac{H_1}{C(S \cap H_1)} \times \dots \times \frac{H_p}{C(S \cap H_p)} .$$

Note that φ is f-holomorphic and

$$F^{\pm} \subset \sum_{i=1}^{p} h_i^{\mathbb{C}},$$

to see that $d\varphi$ takes values in $\underline{\mathbf{k \cap h}}^{\perp}$. Choose $g \in G$ such that $\varphi(M) \ni \pi(1_G)$. An application of Frobenius' Theorem[3] shows that $L_g \circ \varphi$ factors through $\dfrac{K}{K \cap H}$. □

Examples

Some examples of horizontal f-structures on flag manifolds first came to light when flag manifolds were considered as twistor spaces fibering over Hermitian symmetric spaces. (Lemma 5.4 provides the general twistor result.)

(1) [EL3] Let $G = SU(3)$, then the f-structures on the full flag manifold $\dfrac{SU(3)}{S(U(1)^3)}$ given by equation (B) of Theorem 6.13 are precisely the non-integrable almost complex structures.

(2) [B1] Let $G = Sp(2)$, then G has rank 2 and choose I to be the short simple root. In this case equation (B) defines the non-integrable almost complex structures on $\dfrac{Sp(2)}{U(1) \times Sp(1)}$. This manifold fibres homogeneously over $S^4 = \dfrac{Sp(2)}{Sp(1) \times Sp(1)}$ to give the Penrose fibration.

3 [KN] Vol 1, Chap 1.

(3) [BR2] Burstall-Rawnsley show that harmonic maps of S^2 into S^{2n} and $\mathbb{CP}^n$ lift to maps into flags of height not exceeding 2 which fibre over S^{2n}, $\mathbb{CP}^n$ respectively. Further, these lifts are f-holomorphic with respect to an f-structure defined by equation (A) of Theorem 6.13.

(4) Calabi [C] proved that minimal immersions of S^2 into S^{2n} lift to type (A) f-holomorphic maps into $\dfrac{SO(2n+1)}{U(n)}$.

(5) Eells and Wood [EW] showed that linearly full harmonic maps of S^2 into $\mathbb{CP}^n$ lift to type (A) f-holomorphic maps into $\dfrac{SU(n)}{S(U(1)^n)}$.

(6) Bryant [B2] classified the horizontal f-structures on flag manifolds which have trivial zero eigenspace.

Chapter 7
Integrable f-Holomorphic Orbits on Flags

Let H be a connected closed Lie subgroup of a connected compact Lie group G, and equip the G-flag manifold $^{G}/_{C(S)}$ with a horizontal f-structure. We seek to describe the circumstances under which the f-structure induces an integrable complex structure on the orbit of H in $^{G}/_{C(S)}$.

§1. Integrable Orbits Are Hermitian Symmetric

Firstly, let us suppose H, G, C(S) and F have the property above and set $K = C(S) \cap H$. Working infinitesimally:

$$\mathbf{h}^{\mathbb{C}} = \mathbf{k}^{\mathbb{C}} \oplus \mathbf{m}_1 \oplus \mathbf{m}_{-1}, \tag{7.1}$$

where, for our hypotheses to be satisfied

$$\mathbf{m}_1 = \mathbf{h}^{\mathbb{C}} \cap F^+,$$

$$\mathbf{m}_{-1} = \overline{\mathbf{m}}_1 = \mathbf{h}^{\mathbb{C}} \cap F^-$$

and $\mathbf{m}_1$ and $\mathbf{m}_{-1}$ are $\mathbf{k}^{\mathbb{C}}$ invariant.

Furthermore, since F is horizontal

$$[\mathbf{m}_1, \mathbf{m}_{-1}] \subset [F^+, F^-] \subset c(s)^{\mathbb{C}},$$

and thus

$$[\mathbf{m}_1, \mathbf{m}_{-1}] \subset \mathbf{k}^{\mathbb{C}}. \tag{7.2}$$

Integrability of the complex structure determined by setting $\mathbf{m}_1$ to be the (1,0) vectors implies that

$$[\mathbf{m}_1, \mathbf{m}_1] \subset \mathbf{k}^{\mathbb{C}} \oplus \mathbf{m}_1.$$

Claim $F^+ \perp [F^+, F^+]$

Proof. F^+ and hence $[F^+, F^+]$ are $c(\mathbf{s})$ invariant. $c(\mathbf{s})$ contains a maximal torus $\mathbf{t}$. Thus F^+ is an orthogonal direct sum of root spaces for $\mathbf{t}$, while $[F^+, F^+]$ is an orthogonal direct sum of root spaces for $\mathbf{t}$ and a subspace of $\mathbf{t}$.

Let $\mathbf{g}^\alpha \subset [F^+, F^+] \cap F^+$, then $\exists\, \beta, \gamma \in \Delta(\mathbf{g}^{\mathbb{C}}, \mathbf{t}^{\mathbb{C}})$ with $\mathbf{g}^\beta$ and $\mathbf{g}^\gamma$ lying in F^+ such that $\alpha = \beta + \gamma$. This implies that β lies in $[F^+, F^-] \subset c(\mathbf{s})^{\mathbb{C}}$ by horizontality, which is a contradiction. This is sufficient to prove the claim □

We now see that:

$$\mathbf{m}_1 \subset F^+ \perp [F^+, F^+] \supset [\mathbf{m}_1, \mathbf{m}_1]$$

and thus

$$[\mathbf{m}_1, \mathbf{m}_1] \subset \mathbf{k}^{\mathbb{C}}. \qquad (7.3)$$

Let $\mathbf{p} = \mathrm{Re}\,(\mathbf{m}_1 \oplus \mathbf{m}_{-1})$, we have shown that:

$$\mathbf{h} = \mathbf{k} \oplus \mathbf{p}, \ \ [\mathbf{k}, \mathbf{k}] \subset \mathbf{k}, \ \ [\mathbf{p}, \mathbf{p}] \subset \mathbf{k}, \ \ [\mathbf{k}, \mathbf{p}] \subset \mathbf{p}, \qquad (7.4)$$

and $\mathbf{p}^{\mathbb{C}}$ carries a $\mathbf{k}^{\mathbb{C}}$ invariant, integrable complex structure. In other words any such H orbit in $G/_{C(S)}$ is a Hermitian symmetric space.

Of course, each simply connected Hermitian symmetric space is realized as such an integrable orbit, if only by its description as a flag manifold for which the usual complex structure is a horizontal f-structure. Fortunately, more interesting examples occur, to which we now turn.

§2. Orbits in the Full Flag Manifold

Fix a compact simple Lie group G and maximal torus T. For the rest of this chapter we restrict attention to the full flag manifold $^G/_T$. We will need to use the results on roots set out in §2.6. To avoid dealing with degenerate cases we assume:

$$\mathbf{h} \cup \mathbf{t} \text{ generates } \mathbf{g}. \tag{7.5}$$

Theorem 6.13 shows that each example we seek consists of a direct sum of irreducible examples - by which we mean examples such that the f-structure is one of the 'irreducible' types of formula 6.13 (A) and (B). Explicitly, there is a choice of simple roots $\{\alpha_1,...,\alpha_\ell\}$ such that either

$$F_+ = \{\Sigma\, \mathbf{g}^\alpha \mid n(\alpha) = 1\} \tag{6.13A}$$

or

$$\text{rank } G > 1 \text{ and } F_+ = \{\Sigma\, \mathbf{g}^\alpha \mid n(\alpha) \equiv 1 \bmod n(-\alpha_0) + 1\}; \tag{6.13B}$$

where n is the height function defined in §6.1 and $-\alpha_0$ is the highest root.

We choose elements $\{X_\alpha : \alpha \in \Delta(\mathbf{g}^{\mathbb{C}}, \mathbf{t}^{\mathbb{C}})\}$ satisfying the properties of Lemma 2.9.

Note. When G has rank 1, $^G/_T$ is isomorphic to the two sphere with its integrable complex structure. In this case the G orbit on $^G/_T$, i.e. $^G/_T$ itself, provides the only non-trivial example.

We assume that rank $G > 1$ from now on.

Lemma 7.1. Suppose the H orbit on $G/_T$ has integrable complex structure induced by the f-structure of either (6.13A) or (6.13B), and rank $G > 1$. Then $\dim_{\mathbb{C}} \mathbf{m}_1 = 1$.

Proof. In either case,

$$[\mathbf{m}_1, \mathbf{m}_1] \subset [F^+, F^+] \subset \{g^\alpha \mid n(\alpha) \equiv 2 \bmod n(-\alpha_0) + 1\}$$

since rank $G > 1$. This provides a contradiction to equation (7.3) unless $[\mathbf{m}_1, \mathbf{m}_1] = 0$.

Since $\mathbf{t}$ and $\mathbf{h}$ generate $\mathbf{g}$, we may choose $Z \in \mathbf{m}_1$ with the property that

$$Z = \sum_{i=0}^{\ell} \zeta_i X_{\alpha_i}, \tag{7.6}$$

where either

$$\zeta_i \neq 0 \quad i = 0, \ldots, \ell \text{ in case 6.13B,}$$

or

$$\zeta_0 = 0 \text{ and } \zeta_i \neq 0 \quad i = 1, \ldots, \ell \text{ in case 6.13A.}$$

Let

$$X = \sum_{i=0}^{\ell} \xi_i X_{\alpha_i} \in \mathbf{m}_1.$$

Then since $[\mathbf{m}_1, \mathbf{m}_1] = 0$

$$0 = [X, Z] = \sum_{0 \le i < j \le \ell} (\xi_i \zeta_j - \xi_j \zeta_i)[X_{\alpha_i}, X_{\alpha_j}].$$

$$= \sum_{\substack{0 \le i < j \le \ell \\ \alpha_i + \alpha_j \in \Delta(\mathbf{g}^{\mathbb{C}}, \mathbf{t}^{\mathbb{C}})}} (\xi_i \zeta_j - \xi_j \zeta_i)[X_{\alpha_i}, X_{\alpha_j}]$$

since $[\mathbf{g}^\alpha, \mathbf{g}^\beta] = \begin{cases} 0 & \alpha + \beta \text{ not a root} \\ \mathbf{g}^{\alpha+\beta} & \alpha + \beta \text{ is a root} \end{cases}$

The last expression is therefore a sum over non-zero vectors in distinct root spaces and hence a sum over linearly independent vectors, so we see that

$$\xi_i \zeta_j = \xi_j \zeta_i \quad \text{whenever} \quad \alpha_i + \alpha_j \in \Delta(\mathbf{g}^{\mathbb{C}}, \mathbf{t}^{\mathbb{C}}). \tag{7.7}$$

Since $\zeta_i \neq 0$ for the appropriate values of i, we see that

$$\xi_i \zeta_j = \xi_j \zeta_i \text{ and } \xi_j \zeta_k = \xi_k \zeta_j \Rightarrow \xi_i \zeta_k = \xi_k \zeta_i.$$

g is simple, and so the simple root spaces cannot be split into mutually commuting subsets, hence

$$\xi_i \zeta_j = \xi_j \zeta_i \quad \text{for all } i,j.$$

Whence X is a multiple of Z. □

The following corollary follows trivially.

Corollary 7.2. Suppose H satisfies the hypotheses of the Lemma, then there exists $Z \in F_+$ such that:

$$\mathbf{h}^{\mathbb{C}} = \text{linear span of } Z, \bar{Z}, [Z, \bar{Z}]; \tag{7.8}$$

from which it follows that

$$[[Z, \bar{Z}], Z] = c\,Z, \quad c \in \mathbb{C}. \tag{7.9}$$

Conversely, any $Z \in F_+$ which satisfies (7.9) determines, by (7.8), a Lie sub-algebra of **g** satisfying the hypotheses of the Lemma. □

Note that since

$$[Z, \bar{Z}] \in [F_+, F_-] = \mathbf{t}^{\mathbb{C}},$$

equation (7.9) is equivalent to

$$\zeta_i\, \alpha_i([Z, \bar{Z}]) = c\, \zeta_i \qquad i = 0, \ldots, \ell. \tag{7.10}$$

We must now consider separately the cases governed by equations 6.13A and 6.13B.

§3. Case 6.13B. : $F_+ = \{\Sigma\, \mathbf{g}^\alpha : n(\alpha) \equiv 1 \bmod (n(-\alpha_0)+1)\}$.

In this case (7.10) implies

$$\alpha_i([Z, \bar{Z}]) = c \qquad i = 0,\dots,\ell.$$

However,

$$\alpha_0 = -\sum_{i=1}^{\ell} a_i\, \alpha_i$$

(here a_i are the coefficients of the highest root in the basis given by the simple roots), and thus

$$c = \alpha_0([Z, \bar{Z}]) = -\Sigma\, a_i\, \alpha_i([Z, \bar{Z}]) = (-\Sigma\, a_i)c\,.$$

Rank $\mathbf{g} > 1$ so that we must have $c = 0$ for the equation above to hold. But $\{\alpha_i \;\; i = 1,\dots,\ell\}$ span the dual of $\mathbf{t}^{\mathbb{C}}$, so:

$$[Z, \bar{Z}] = 0. \tag{7.11}$$

Lastly, we solve the equation

$$0 = [Z, \bar{Z}] = \sum_{i=0}^{\ell} \zeta_i \bar{\zeta}_i H_{\alpha_i}\,; \;\; Z \in F^+. \tag{7.12}$$

Now, H_{α_i} $i = 1,\dots,\ell$ are linearly independent and

$$H_{\alpha_0} = -\sum_{i=1}^{\ell} a_i\, H_{\alpha_i}.$$

So in order that (11) holds

$$\zeta_i = \sqrt{a_i} \cdot u_i \qquad\qquad 1 \le i \le \ell,$$

where u_i is a unit complex number, and choosing the normalization $|\zeta_0| = 1$.

Thus the solutions we seek are:

$$Z = u_0 X_{\alpha_0} + \sum_{i=1}^{\ell} u_i \sqrt{a_i} X_{\alpha_i} \qquad u_i \in S^1 \subset \mathbb{C}. \tag{7.13}$$

The sub-algebra of $\mathbf{g}^{\mathbb{C}}$ generated by Z and $\bar{Z}$ is a two complex dimensional toral sub-algebra, and hence its real part $\mathbf{h}$ is a two real dimensional toral sub-algebra of $\mathbf{g}$. The various possible choices of $\mathbf{h}$ are parameterized by the choice of the u_i $i = 0,\ldots,\ell$ of equation (7.13) modulo multiplication of Z by a unit complex constant. Thus we may parameterize family of such toral subalgebras $\mathbf{h}$ by T, the maximal torus of G.

Notice that the isotropy of the action of H on G/T is discrete and hence the orbit of H in G/T is also a 2-torus.

§4. Case 6.13A. : $F_+ = \{\Sigma \mathbf{g}^{\alpha} : n(\alpha) = 1\}$.

In this case equation (7.10) implies

$$\alpha_i([Z, \bar{Z}]) = c \quad i = 1,\ldots,\ell. \tag{7.14}$$

Lemma 7.3. (c.f. [A] Prop. 5.4.2)

Let $H = \sum_{\alpha \text{ positive}} \frac{1}{\langle\alpha,\alpha\rangle} H_\alpha$, then $\alpha_i(H) = 1$ for all simple roots α_i.

Proof. Let α_i be a simple root, we can define an element φ_i of the Weyl group by:

$$\varphi_i(\alpha) = \alpha - \frac{2\langle\alpha_i, \alpha\rangle}{\langle\alpha_i, \alpha_i\rangle}\alpha_i.$$

φ_i has the property that it permutes positive roots other than α_i and sends α_i to $-\alpha_i$ (see [A] for instance).

There are two cases to consider-

(i) φ_i permutes the positive roots $\beta_1,\ldots,\beta_n$ transitively. Then since the Weyl group acts by isometries on roots, the β_j have the same length and

$$\langle\alpha_i, \beta_1 + \ldots + \beta_n\rangle = 0$$

because φ_i fixes the weight $\sum_{j=1}^{n} \beta_j$. Thus $\sum_{j=1}^{n} \frac{1}{\langle\beta_j, \beta_j\rangle}\alpha_i(H_{\beta_j}) = 0.$

(ii) $\frac{1}{\langle\alpha_i, \alpha_i\rangle}\alpha_i(H_{\alpha_i}) = 1.$

Thus $\alpha_i(H) = 1.$ □

Because α_i $i = 1,\ldots,\ell$ span the dual space of $\mathbf{t}^{\mathbb{C}}$, Lemma 7.3 shows that $v \in \mathbf{t}^{\mathbb{C}}$ is a solution to

$$\alpha_i(v) = c \quad i = 1,\ldots,\ell$$

precisely when $v = cH$.

On the other hand,

$$[Z, \bar{Z}] = \sum_{i=1}^{\ell} \zeta_i \bar{\zeta}_i H_{\alpha_i}$$

lies in the subset P of $t^{\mathbb{C}}$ generated by strictly positive real linear combinations of the basis $\{H_{\alpha_i}, \ldots, H_{\alpha_\ell}\}$ for $t^{\mathbb{C}}$. Suppose α is a positive root, then

$$\alpha = \Sigma\, n_i \alpha_i \quad n_i \geq 0,$$

and dually $H_\alpha = \Sigma\, n_i H_{\alpha_i}, \; n_i \geq 0.$

Thus $cH \in P \Leftrightarrow c > 0.$

Finally, we see that the possible choices for Z are parameterised by $U(1)^{\ell} \times \mathbb{R}^+$ and the sub-algebras which Z generates are parameterized by $U(1)^{\ell-1}$. Hence the sub-algebras **h** we have constructed are isomorphic to su(2) and parameterized by $U(1)^{\ell-1}$. The orbit of H on $^G/_T$ is isomorphic to a two sphere carrying its integrable holomorphic structure.

§5 Concluding Remarks

In this Chapter we have obtained examples of horizontal holomorphic and hence equi-harmonic maps. Further examples of horizontal holomorphic maps may be generated by pre-composition of a holomorphic map with the equivariant examples above. On the other hand, additional examples of equi-harmonic (but not neccessarily f-holomorphic) maps may be produced by post composition of a homogeneous projection to a horizontal holomorphic map (Lemma 5.4).

It would be interesting to identify examples of integrable f-holomorphic orbits on flag manifolds other than the full flag manifold; however, a better technique for handling the algebra seems to be a prerequisite.

Chapter 8
Equi-Minimal Maps of Riemann Surfaces to Full Flag Manifolds

On a two dimensional domain the energy functional (1.1) is conformally invariant. Thus on a Riemann surface the harmonicity of a map depends only on the complex structure and the Euler–Lagrange equations for harmonic maps of Riemann surfaces can be reformulated to reflect this fact. Let (U, z) be a holomorphic coordinate chart on a Riemann surface M, then $\varphi : (M, J) \to (N, h)$ is harmonic iff

$$\nabla_{\bar{z}}\, \varphi_*\left(\frac{\partial}{\partial z}\right) = 0, \tag{8.1}$$

where ∇ is the connection on $\varphi^{-1}\, T^{\mathbb{C}}N$. Koszul and Malgrange [KM] proved that any connection ∇ on a complex vector bundle over a Riemann surface has a unique compatible holomorphic structure, such that a local section σ over (U, z) is holomorphic iff $\nabla_{\bar{z}}\sigma = 0$. We can thus re-interpret equation (1) as stating that φ is harmonic exactly when $d\varphi^{1,0}$ is holomorphic section of $T^{1,0}_{*} M \otimes \varphi^{-1}T^{\mathbb{C}}N$.

The map φ is said to be *weakly conformal* if

$$\varphi^* h^{2,0} = 0, \tag{8.2}$$

where h is the complex bilinear extension of the metric on N. Maps which are harmonic and weakly conformal are branched minimal immersions in the sense of Gulliver-Osserman-Royden. Furthermore, when φ is harmonic $\varphi^*h^{2,0}$ is a holomorphic quadratic differential on M, and thus if M has genus zero $\varphi^*h^{2,0}$ must vanish. In other words, a harmonic map of S^2 is automatically weakly conformal.[1]

§1. Equi-Minimal Maps are Horizontal Holomorphic

Definition. A map into a homogeneous space $G/_H$ with non-empty set of G invariant metrics will be said to be *equi-minimal* (respectively *equi-weakly conformal*) if φ is minimal (respectively weakly conformal) with respect to each of the G invariant metrics on $G/_H$.

Let G be a compact semi-simple Lie group with maximal torus T. The homogeneous space we shall be considering in this chapter is $G/_T$, the *full flag manifold* of G. As is traditional, φ denotes a smooth mapping of Riemann surface (M, J) into $G/_T$.

Let $x \in M$. At $\varphi(x)$ we have the (unique) reductive splitting

$$\mathbf{g}^{\mathbb{C}} = \mathbf{t}^{\mathbb{C}}_{\varphi(x)} \oplus \mathbf{m}^{\mathbb{C}}_{\varphi(x)}, \quad \mathbf{m}^{\mathbb{C}}_{\varphi(x)} = \bigoplus_{\alpha \in \Delta} \mathbf{m}_\alpha ;$$

where $\Delta = \Delta(\mathbf{g}^{\mathbb{C}}, \mathbf{t}^{\mathbb{C}})$. Let (U, z) be a holomorphic chart for M containing x, whence

$$\varphi_*(\frac{\partial}{\partial z}(x)) \in \mathbf{m}^{\mathbb{C}}_{\varphi(x)} .$$

Choose a basis $\{X_\alpha : \alpha \in \Delta\}$ for $\mathbf{m}^{\mathbb{C}}_{\varphi(x)}$ in accordance with the properties of Lemma 2.9. Write

$$\varphi_*(\frac{\partial}{\partial z}(x)) = \sum_{\alpha \in \Delta} \zeta_\alpha X_\alpha, \tag{8.3}$$

and note that using Lemma 2.9 :

$$\varphi_*(\frac{\partial}{\partial \bar{z}}(x)) = \overline{\varphi_*(\frac{\partial}{\partial z}(x))} = \sum_{\alpha \in \Delta} \overline{\zeta_{-\alpha}} X_\alpha. \tag{8.4}$$

---

[1] See [EL1] for proofs and further details relating to this paragraph.

Lemma 8.1. φ is equi-weakly conformal at $x \in M$ iff it is almost holomorphic at x

i.e.

$$\zeta_\alpha \neq 0 \Rightarrow \zeta_{-\alpha} = 0.$$

Proof. Use equation (8.2) and formula 2.4 to show that equi-weak conformality is equivalent to:

$$0 = \sum_{\alpha \in \Delta} \lambda_\alpha B(\zeta_\alpha X_\alpha + \zeta_{-\alpha} X_{-\alpha}, \zeta_\alpha X_\alpha + \zeta_{-\alpha} X_{-\alpha}) \tag{8.5}$$

for all choices of $\lambda_\alpha > 0$. B denotes the Killing form on $\mathbf{g}^{\mathbb{C}}$, $\mathbf{g}$ is semi-simple so that B restricted to $\mathbf{g}$ is a bi-invariant metric. Taking linear combinations of (8.5) over appropriate choices of λ_α we see that (8.5) is equivalent to:

$$\begin{aligned} 0 &= B(\zeta_\alpha X_\alpha + \zeta_{-\alpha} X_{-\alpha}, \zeta_\alpha X_\alpha + \zeta_{-\alpha} X_\alpha) && \forall \alpha \in \Delta, \\ &= 2\,\zeta_\alpha \zeta_{-\alpha} && \forall \alpha \in \Delta, \end{aligned}$$

using the special properties of the basis X_α.

This proves the result. □

Recall the notation of Section 2.4, i.e. let β denote the Maurer–Cartan form of ${}^{G}/_{T}$ and P_α the projection onto the irreducible summand of $T^{G}/_{T}$ indexed by α.

Lemma 8.2. If φ is equi-minimal, then

(a) $\varphi^* [\beta \wedge \beta] \subset \mathbf{t}$,

(b) $\varphi^* [\beta \wedge (P_\alpha - P_{-\alpha})\beta] = 0 \quad \forall \alpha \in \Delta$,

(c) $\varphi^* \beta_\alpha^{1,0}$ is a holomorphic section of $T^*_{1,0} M \otimes \varphi^{-1} \mathbf{m}_\alpha$.

Proof. Since $\mathrm{Re}(\mathbf{m}_\alpha \oplus \mathbf{m}_{-\alpha})$ are the irreducible T spaces of **m**, the equi-harmonic version of Nöther's Theorem (Theorem 5.2) proves that since φ is equi-harmonic

$$d^*(\varphi^* \beta_\alpha + \varphi^*\beta_{-\alpha}) = 0 \quad \forall \alpha \in \Delta. \tag{8.6}$$

Also,

$$\begin{aligned} d^*(\varphi^*\beta_\alpha + \varphi^*\beta_{-\alpha}) &= * \, d * (\varphi^*\beta_\alpha + \varphi^* \beta_{-\alpha}) \\ &= - * d(\varphi^* \beta_\alpha \circ J + \varphi^* \beta_{-\alpha} \circ J). \end{aligned} \tag{8.7}$$

because (M, J) is a Riemann surface.

Suppose that $\varphi^* \beta_\alpha^{1,0}$ is non-zero on an open set V in M, and apply Lemma 8.1 to the points in V to see that $\varphi^* \beta_\alpha^{0,1} = 0$ on V, and hence, by conjugation $\varphi^*\beta_{-\alpha}^{1,0} = 0$ on V.

In this case, we have

$$\varphi^*\beta_\alpha \circ J + \varphi^*\beta_{-\alpha} \circ J = i(\varphi^*\beta_\alpha - \varphi^*\beta_{-\alpha}).$$

On the other hand, if $\varphi^* \beta_\alpha^{0,1}$ is non-zero then by a similar argument:

$$\varphi^*\beta_\alpha \circ J + \varphi^* \beta_{-\alpha} \circ J = - i(\varphi^* \beta_\alpha - \varphi^* \beta_{-\alpha}).$$

In either case (8.6) and (8.7) imply that

$$d\varphi^*\beta_\alpha = d\varphi^*\beta_{-\alpha} \qquad \forall \alpha \in \Delta. \tag{8.8}$$

We know from §2.4 that

$$d\beta_\alpha = [\, \beta \wedge \beta_\alpha \,] - \tfrac{1}{2} P_\alpha [\, \beta \wedge \beta \,],$$

so (8.8) may be written as

$$\varphi^* \{ [\, \beta \wedge (P_\alpha - P_{-\alpha})\beta \,] - \tfrac{1}{2} (P_\alpha - P_{-\alpha}) [\, \beta \wedge \beta \,] \} = 0 \qquad \forall \alpha \in \Delta. \tag{8.9}$$

Consider the component of (8.9) in $\mathbf{m}_\alpha$. The contribution from the first summand is:

$$\varphi^* [\beta_\gamma \wedge \beta_{-\alpha}],$$

where γ is a root such that

$$\gamma - \alpha = \alpha, \text{ i.e. } \gamma = 2\alpha.$$

Root systems for Lie groups contain no multiple roots (Theorem 2.8) so that no such γ exists, hence

$$P_\alpha[\beta \wedge (P_\alpha - P_{-\alpha})\beta] = 0.$$

Thus equation (8.9) is equivalent to

$$\varphi^*[\beta \wedge (P_\alpha - P_{-\alpha})\beta] = 0, \quad \forall \alpha \in \Delta; \tag{8.9'}$$

and

$$\varphi^*(P_\alpha - P_{-\alpha})[\beta \wedge \beta] = 0, \quad \forall \alpha \in \Delta; \tag{8.9''}$$

which yield formulae (a) and (b).

The map φ is equi-minimal, so in particular φ is harmonic with respect to the reductive metric on G/T. Let ∇^R, ∇^c denote the Levi-Civita connection of this metric and the canonical connection respectively. On a holomorphic chart in Riemann surface M, the Euler-Lagrange equation for φ may be written as

$$\nabla^R_{\bar{z}} \varphi^*\beta\left(\frac{\partial}{\partial z}\right) = 0$$

Recall from §2.5 that

$$\nabla^R_{\bar{z}} \varphi^*\beta\left(\frac{\partial}{\partial z}\right) = \nabla^c_{\bar{z}} \varphi^* \beta\left(\frac{\partial}{\partial z}\right) + \tfrac{1}{2}P_{\mathbf{m}}[\varphi^*\beta\left(\frac{\partial}{\partial \bar{z}}\right), \varphi^*\beta\left(\frac{\partial}{\partial z}\right)].$$

Equation (a) shows that in our case the last term vanishes, hence

$$\nabla^c_{\bar{z}} \varphi^*\beta\left(\frac{\partial}{\partial z}\right) = 0. \tag{8.10}$$

The sub-bundles $\mathfrak{m}_\alpha$ of $\mathfrak{m}$ are parallel for the canonical connection so that

$$\nabla^c_{\bar{z}}\, \varphi^* \beta_\alpha \left(\frac{\partial}{\partial z} \right) = 0, \qquad \forall \alpha \in \Delta. \tag{8.11}$$

Equip $\varphi^{-1}\mathfrak{m}_\alpha$ with the Koszul-Molgrange holomorphic structure induced by the canonical connection restricted to $\mathfrak{m}_\alpha$. With respect to this structure, statement (c) is the global version of equation (8.11). □

We now come to the main result of this chapter.

Theorem 8.3. An equi-minimal map of a Riemann surface into a full flag manifold is horizontal holomorphic.

Proof. Let $\varphi : M \to {}^G/_T$ be such an equi-minimal map. An immediate consequence of part (c) of Lemma 8.2 is that since φ is equi-minimal the forms $\varphi^*\beta_\alpha^{1,0}$ vanish either identically, or only on an isolated set of points. We can therefore choose a point $x \in M$ which is generic in the sense that

$$(\varphi^* \beta_\alpha^{1,0})_x = 0 \Leftrightarrow \varphi^* \beta_\alpha^{1,0} \equiv 0 \qquad \forall \alpha \in \Delta.$$

Recall the methods of Section 4.1, and use the notation of equation 8.3 to define

$$F_+ = \{\alpha \in \Delta \mid \zeta_\alpha = 0\}.$$

Lemma 8.1 proves that $F_+ \cap \overline{F_+} = \{0\}$, so we see that F_+ defines an invariant f-structure on ${}^G/_T$. We are working at a generic point x and so φ is (globally) f-holomorphic with respect to this f-structure. It only remains to establish that this f-structure is horizontal.

Stated in terms of roots, the condition that F_+ be horizontal is (c.f. §4.2 and §2.6):

$$\alpha, \beta \in F_+ \Rightarrow \alpha - \beta \notin \Delta.$$

Equivalently:

$$\alpha, \beta, \alpha - \beta \in \Delta; \ \alpha \in F_+ \Rightarrow \beta \notin F_+.$$

Set $\gamma = \beta - \alpha$ and we can rewrite this as:

$$\alpha, \gamma, \alpha + \gamma \in \Delta; \ \alpha \in F_+ \Rightarrow \alpha + \gamma \notin F_+. \tag{8.12}$$

Now apply the definition of F_+ to obtain:

F_+ is horizontal iff whenever $\alpha, \gamma, \alpha + \gamma$ are roots such that (8.13)

$\zeta_\alpha \neq 0$ then $\zeta_{\alpha+\gamma} = 0$.

The formulae which follow will be simplified by adopting the convention that P_δ and ζ_δ are zero when δ is not a root and $N_{\alpha,\beta}$ is zero if any one of α, β or $\alpha + \beta$ fails to be a root. Note that $N_{\alpha,\beta}$ is non-zero unless this convention is in force (see Lemma 2.9 for the definition of the numbers $N_{\alpha,\beta}$).

The component of equation (b) of Lemma 8.2 lying in $\mathbf{m}_\gamma$ is

$$\begin{aligned} 0 &= \varphi^* P_\gamma [\beta \wedge (P_\alpha - P_{-\alpha})\beta] \\ &= \varphi^* [P_{\gamma-\alpha} \beta \wedge P_\alpha \beta] - \varphi^* [P_{\gamma+\alpha} \beta \wedge P_{-\alpha} \beta]. \end{aligned}$$

Evaluate these forms on $\dfrac{\partial}{\partial z} \wedge \dfrac{\partial}{\partial \bar{z}}$ at x to obtain:

$$\begin{aligned} 0 = {} & [\zeta_{\gamma-\alpha} X_{\gamma-\alpha}, \overline{\zeta_{-\alpha}} X_\alpha] - [\overline{\zeta_{\alpha-\gamma}} X_{\gamma-\alpha}, \zeta_\alpha X_\alpha] \\ & - [\zeta_{\gamma-\alpha} X_{\gamma+\alpha}, \overline{\zeta_\alpha} X_{-\alpha}] + [\overline{\zeta_{-\gamma-\alpha}} X_{\gamma+\alpha}, \zeta_{-\alpha} X_{-\alpha}]. \end{aligned}$$

However, Lemma 8.1 shows that either ζ_α or $\zeta_{-\alpha}$ is zero, so we have:

$$0 = [\overline{\zeta_{\alpha-\gamma}}\, X_{\gamma-\alpha}, \zeta_\alpha X_\alpha] + [\zeta_{\gamma+\alpha} X_{\gamma+\alpha}, \overline{\zeta_\alpha}\, X_{-\alpha}].$$

We can rewrite this as

$$0 = \overline{\zeta_{\alpha-\gamma}}\, \zeta_\alpha N_{\gamma-\alpha,\alpha} + \zeta_{\gamma+\alpha} \overline{\zeta_\alpha} N_{\gamma+\alpha,-\alpha} \qquad \forall \alpha, \gamma \in \Delta. \qquad (8.14)$$

Claim. If $\alpha, \gamma, \alpha + \gamma \in \Delta$ then $\zeta_\alpha \neq 0 \Rightarrow \zeta_{\gamma+\alpha} = 0$.

The proof of this claim is inductive.

(i) Suppose $\alpha, \gamma, \gamma + \alpha \in \Delta$; $\gamma - \alpha \notin \Delta$ and $\zeta_\alpha \neq 0$. $\gamma-\alpha$ is not a root, so $N_{\gamma-\alpha,\alpha} = 0$. On the other hand $N_{\gamma+\alpha,-\alpha} \neq 0$. (8.13) reduces to

$$\zeta_{\gamma+\alpha} \overline{\zeta_\alpha} = 0.$$

Hence $\zeta_{\gamma+\alpha} = 0$.

(ii) Suppose $\alpha, \gamma, \alpha + \gamma, \alpha - \gamma \in \Delta$; $\alpha - 2\gamma \notin \Delta$; $\zeta_\alpha \neq 0$.

If $\zeta_{\alpha+\gamma} \neq 0$, then in order that (8.14) be satisfied we must have $\zeta_{\alpha-\gamma} \neq 0$. We can apply (i) to the roots $\alpha - \gamma$ and γ to conclude that $\zeta_\alpha = 0$ which is a contradiction. Thus $\zeta_{\alpha+\gamma} = 0$.

(iii) Suppose $\alpha, \gamma, \alpha + \gamma, \alpha - \gamma, \alpha - 2\gamma \in \Delta$; $\alpha - 3\gamma \notin \Delta$; $\zeta_\alpha \neq 0$.

If $\zeta_{\alpha+\gamma} \neq 0$ again we must have $\zeta_{\alpha-\gamma} \neq 0$. Apply (ii) to the roots $\alpha - \gamma$ and γ to conclude once again $\zeta_\alpha = 0$. The contradiction proves that $\zeta_{\alpha+\gamma} = 0$.

The maximal length of root strings in 4 (Theorem 2.8), so we have now proved the claim and, by invoking (8.12), the theorem. □

§2. Branched Horizontal Curves In Full Flags

Theorem 6.13 shows that the irreducible f-structures on full flag manifolds take only two possible forms:

(A) A set of simple roots

or

(B) A set of simple roots and minus the highest root (this case can only occur when rank $G > 1$).

Notation. A mapping of a Riemann surface which is f-holomorphic with respect to an f-structure of type (A) (respectively (B)) will be called a *branched horizontal curve of type* (A) (respectively, type (B)).

Theorem 8.4. Let $\varphi : M \to {}^{G}/_{T}$. Suppose M is connected and $\varphi(M) \ni \pi(1_G)$. Then φ is equi-minimal iff there exists a set H_i $(i = 1,...,p)$ of mutually orthogonal, commuting, simple Lie subgroups of G such that φ factors as

$$\varphi : M \to {}^{H_1}/_{T_1} \times \ldots \times {}^{H_p}/_{T_p} \hookrightarrow {}^{G}/_{T},$$

and such that the ith component of φ is a full branched horizontal curve in ${}^{H_i}/_{T_i}$.

Note. $T_i = T \cap H_i$ is a maximal torus for H_i.

Proof. To prove necessity apply Corollary 6.14 and Theorem 8.3. To prove sufficiency use Corollary 4.4 and Lemma 8.1.

When the Riemann surface M is equal to S^2, the situation simplifies further.

Theorem 8.5. Let $\varphi : S^2 \to G/T$ be a full branched horizontal curve, then φ is of type (A).

Proof. Suppose φ is of type (B) and let $\alpha_1,\ldots,\alpha_\ell$ be the set of simple roots such that φ is f-holomorphic with respect to f-structure (B). Let α_0 denote minus the highest root. Lemma 8.2c) proves that we have sections $\varphi^* \beta^{1,0}_{\alpha_i}$ $(i = 0,\ldots,\ell)$ of $T^*_{1,0}M \otimes \varphi^{-1}\underline{\mathbf{m}}_\alpha$ which are holomorphic with respect to the holomorphic structure induced by the canonical connection. The map φ is full and so none of these sections vanishes identically.

Use Lemma 2.10 to express the highest root as

$$-\alpha_0 = \alpha_{i_1} + \ldots + \alpha_{i_r}, \qquad 1 \leq i_j \leq \ell;$$

such that each partial sum of the form

$$\alpha_{i_1} + \ldots + \alpha_{i_s}, \qquad 1 \leq s \leq r$$

is a root.

Lie bracket is a G invariant tensor on G/T and hence parallel for the canonical connection, we can thus construct a holomorphic differential

$$\gamma = [\varphi^*\beta^{1,0}_{\alpha_0}, [\varphi^* \beta^{1,0}_{\alpha_{i_r}}, [\ldots, \varphi^* \beta^{1,0}_{\alpha_{i_1}}] \ldots].$$

Now, γ is a holomorphic section of $\otimes^{r+1} T^*_{1,0} S^2 \otimes \mathfrak{t}^{\mathbb{C}}$ which is not identically zero, where the holomorphic structure of $\mathfrak{t}^{\mathbb{C}}$ is induced by the canonical connection restricted to $\mathfrak{t}^{\mathbb{C}}$.

The elements H_α defined in Theorem 2.7 define global sections for $\mathfrak{t}^{\mathbb{C}}$. The H_α are invariant and hence covariant constant for the canonical connection, so we see that the canonical connection restricted to $\mathfrak{t}^{\mathbb{C}}$ coincides with the flat connection induced by this trivialization of $\mathfrak{t}^{\mathbb{C}}$.

Hence γ is a non-zero holomorphic differential on S^2, which is a contradiction. □

Branched Horizontal Curves of Type (A)

These curves are holomorphic with respect to an integrable complex structure on G/T and as such may be investigated by using the techniques of complex analysis and algebraic geometry.

When $G = SU(n)$, we can choose a homogeneous projection π of G/T to $\mathbb{CP}^{n-1}$ with the property that if φ is a branched horizontal curve of type (A) then $\pi \circ \varphi$ is a linearly full holomorphic map into $\mathbb{CP}^{n-1}$. The branched horizontal curve may then be recovered by using the associated curves of $\pi \circ \varphi$. This procedure yields a bijective correspondence between such curves and linearly full holomorphic maps of Riemann surfaces into complex projective spaces - the study of which constitutes a rich and venerable part of mathematics. [GH §2.4] provides an exposition on this topic.

It is interesting to note that Eells and Wood [EW] established that a map of S^2 into $\mathbb{CP}^{n-1}$ is linearly full and harmonic precisely when it is the image, under homogeneous projection, of a branched horizontal curve of type (A) in $SU(n)/T$. The paper [EW] prompted a study of harmonic maps into Grassmanians (see for Example [BW], [CW], [W]) from which it is clear that no such simple lifting exists for Grassmanians. Negreiros [N1] proved that such curves into $\mathbb{CP}^{n-1}$ (which he calls 'Eells - Wood' maps) are equi-harmonic and calculated for which metrics they are stable.

Yang [Yang 1], [Yang 2] studied branched horizontal curves of type (A) when $G = Sp(n)$. In analogy to the $SU(n)$ case, he was able to establish Plücker type formulae. Yang also considered horizontal curves of type (A) without branch points (i.e. immersions) and obtained 'quantization' results: i.e. restrictions on the Gaussian curvature of the pull back metric on M.

Bryant [B2] proved that a type (A) curve into any compact group G composed with a homogeneous projection onto a Hermitian symmetric space is a harmonic map. Indeed, this result seems to have been the motivation for Yang's work.

Branched Horizontal Curves of Type (B)

In contrast to the previous case, little is known concerning curves of type (B). Indeed the only examples I know of are those constructed in Chapter 7 of this thesis. The reason for the sparsity of examples in this case is the fact that curves of type (B) are pseudo-holomorphic (i.e. holomorphic with respect to a non-integrable almost complex structure). Recently however, progress has begun to be made in the study of pseudo holomorphic curves, for example see [Gr].

References

[A] Adams, J.F.: Lectures on Lie Groups. Benjamin, New York, 1969.

[B1] Bryant, R.: Conformal and minimal immersions of compact surfaces into the 4-sphere. J. Diff. Geom. 17 (1982) 455-473.

[B2] Bryant, R.: Lie groups and twistor spaces. Duke Math. J. 52 (1985) 223-261.

[BH] Borel, A. and Hirzebruch, F.: Characteristic classes and homogeneous spaces, I. Amer. J. Math. 80 (1958) 459-538.

[BR1] Burstall, F.E. and Rawnsley, J.H.: Sphères harmoniques dans les groupes de Lie compacts et courbes holomorphes dans les espaces homogènes. C.R. Acad. Sci. Paris 302 (1986) 709-712.

[BR2] Burstall, F.E. and Rawnsley, J.H.: Twistor theory for Riemannian symmetric spaces. Lecture Notes in Math. 1424. Springer, Berlin, 1990.

[BS] Burstall, F.E. and Salamon, S.M.: Tournaments, flags and harmonic maps. Math. Ann. 277 (1987) 249-265.

[BW] Burstall, F.E. and Wood, J.C.: The construction of harmonic maps into complex Grassmanians. J. Diff. Geom. 23 (1986) 255-297.

[C] Calabi, E.: Minimal immersions of surfaces in Euclidean spheres. J. Diff. Geom. 1 (1967) 111-125.

[CW] Chern, S.S. and Wolfson, J.: Harmonic maps of the two-sphere into a complex Grassmanian manifold, II. Ann. of Math. 125 (1987) 301-335.

[D] Ding, W.Y. and Chen, Y.M.: Blow up and global existence of heat flow of harmonic maps. Invent. Math. 99 (1990) 567-578.

[EL1] Eells, J. and Lemaire, L.: A report on harmonic maps. Bull. Lond. Math. Soc. 10 (1978) 1-68.

[EL2] Eells, J. and Lemaire, L.: Selected topics in harmonic maps. C.B.M.S. Regional Conf. Ser. 50 in Math. Providence, RI: Amer. Math. Soc. 1983.

[EL3] Eells, J. and Lemaire, L.: Another report on harmonic maps. Bull. Lond. Math. Soc. 20 (1988) 385-524.

[ES] Eells, J. and Sampson, J.H.: Harmonic mappings of Riemannian manifolds. Amer. J. Math. 86 (1964) 109-160.

[EW] Eells, J. and Wood, J.C.: Harmonic maps from surfaces into projective spaces. Adv. in Math. 49 (1983) 217-263.

[Gr] Gromov, M.: Pseudo holomorphic curves in symplectic manifolds, Invent. Math. 82 (1985) 307-347.

[Gu] Guest, M.: Geometry of maps between generalised flag manifolds. J. Diff. Geom. 25 (1987) 223-247.

[GH] Griffiths, P. and Harris, J.: Principles of Algebraic Geometry, Wiley, New York, 1978.

[He] Helgason, S.: Differential Geometry, Lie Groups, and Symmetric Spaces. Academic Press, New York, San Francisco, London, 1978.

[Hu] Humphreys, J.E.: Introduction to Lie Algebras and Representation Theory. Springer, New York, Heidelberg, London, 1972.

[HL] Harvey, R. and Lawson, B.: Calibrated geometries. Acta Math. 148 (1982) 47-157.

[IM] Itoh, M. and Manabe, H.: Yang-Mills-Higgs Fields and Harmonicity of Limit Maps. Proc. Japan Acad., 65, Ser. A (1989) 113-115.

[J] Jacobson, N.: Lie Algebras. Wiley (Interscience), New York, 1962.

[KN] Kobayashi, S. and Nomizu, K.: Foundations of Differential Geometry I, II. Wiley (Interscience), New York, 1963, and 1969.

[L] Lichnerowicz, A.: Applications harmoniques et variétés kählériennes. Sympos. Math. 3 (1970) 341-402.

[La] Lawson, H. B.: Lectures on Minimal Submanifolds. Second edition, Publish or Perish, 1980.

[LKV] Le Khong Van.: Minimal surfaces in homogeneous spaces. Math. USSR Izvestiya 32 (1989) 413-427.

[N1] Negreiros, C.J.C.: Harmonic maps from compact Riemann surfaces into flag manifolds. Thesis, University of Chicago, 1987.

[N2] Negreiros, C.J.C.: Some remarks about harmonic maps into flag manifolds. Preprint, Univrsidade Estadual de Campinas, 1987.

[N3] Negreiros, C.J.C.: Equivariant harmonic maps into flag manifolds. Preprint, Universidade Estadual de Campinas, 1987.

[P] Pluzhnikov, A.I.: Some properties of harmonic mappings in the case of spheres and Lie groups. Sov. Math. Dokl. 27 (1983) 246-248.

[R1] Rawnsley, J.H.: Nöther's theorem for harmonic maps. Diff. Geom. Methods in Math. Phys. Reidel, 1984 pp. 197–202.

[R2] Rawnsley, J.H.: f-structures, f-twistor spaces and harmonic maps. Sem. Geom. L. Bianchi II 1984. Lecture Notes in Math. 1164, Springer, Berlin 1985 pp. 85–159.

[S] Salamon, S.M.: Harmonic and holomorphic maps. Sem. Geom. L. Bianchi II 1984. Lecture Notes in Math. 1164. Springer, Berlin 1985. pp. 161–224.

[St] Struwe, M.: On the evolution of harmonic mappings of Riemannian surfaces. Comment. Math. Helv. 60 (1985) 558–581.

[U] Uhlenbeck, K.: Harmonic maps into Lie groups (classical solutions of the chiral model). J. Diff. Geom. 30 (1989) 1–50.

[V] Valli, G.: On the energy spectrtum of harmonic 2-spheres in unitary groups. Topology 27 (1988) 129–136.

[Wa] Warner, F.: Foundations of Differentiable Manifolds and Lie Groups. Graduate Texts in Mathematics 94, Springer-Verlag, New York, 1983.

[Wolf] Wolf, J.A.: Spaces of constant curvature. McGraw-Hill, New York, 1967.

[Wood] Wood, J.C.: The explicit construction and parametrization of all harmonic maps from the two sphere to a complex Grassmanian. J. Reine Angew. Math. 386 (1988) 1-31.

[Yang 1] Yang, K.: Almost complex homogeneous spaces and their submanifolds. World Scientific, Singapore, 1987.

[Yang 2] Yang, K.: Horizontal holomorphic curves in Sp(n)-flag manifolds. Proc. A.M.S. 103 (1988) 265–273.

[Yano] Yano, K.: On a structure defined by a tensor field of type (1,1) satisfying $F^3 + F = 0$. Tensor 14 (1963) 99–109.